To Feel What Others Feel

Social Sources of the Placebo Effect

Perspectives in Medical Humanities

Perspectives in Medical Humanities publishes peer reviewed scholarship produced or reviewed under the auspices of the University of California Medical Humanities Consortium, a multi-campus collaborative of faculty, students, and trainees in the humanities, medicine, and health sciences. Our series invites scholars from the humanities and health care professions to share narratives and analysis on health, healing, and the contexts of our beliefs and practices that impact biomedical inquiry.

General Editor

Brian Dolan, PhD, Professor of Social Medicine and Medical Humanities, University of California, San Francisco (UCSF)

Recent Titles

Clowns and Jokers Can Heal Us: Comedy and Medicine
By Albert Howard Carter III (Fall 2011)

The Remarkables: Endocrine Abnormalities in Art
By Carol Clark and Orlo Clark (Winter 2011)

Health Citizenship: Essays in Social Medicine and Biomedical Politics
By Dorothy Porter (Winter 2011)

What to Read on Love, not Sex: Freud, Fiction, and the Articulation of Truth in Modern Psychological Science
By Edison Miyawaki, MD; Foreword by Harold Bloom (Fall 2012)

Patient Poets: Illness from Inside Out
By Marilyn Chandler McEntyre (Fall 2012)

mh

www.medicalhumanities.ucsf.edu

This series is made possible by the generous support of the Dean of the School of Medicine at UCSF, the Center for Humanities and Health Sciences at UCSF, and a Multicampus Research Program Grant from the University of California Office of the President.

To My Wife and Children

To Feel What Others Feel

Social Sources of the Placebo Effect

Stewart Justman

University of California Medical Humanities Press
2012

First published in 2012
by University of California Medical Humanities Press
in partnership with eScholarship | University of California
San Francisco – berkeley – london

© 2012 by Stewart Justman
University of California
Medical Humanities Consortium
3333 California Street, Suite 485
San Francisco, CA 94143-0850

Cover Design by Eduardo de Ugarte

Library of Congress Control Number: 2012953399

isbn (paperback) 978-0-9834639-9-3

Printed in usa

Contents

Acknowledgments

My gratitude goes to Linda Frey, Marsha Frey, Cory Harris, Richard McNally, Michael Mayer, and Amir Raz; to Drs. Barnett Kramer, Andrew Leuchter, Jean-Luc Mommaerts, Paul Schellhammer, and Anthony Zietman; to Frederick Crews, who first got me interested in the placebo effect; and to series editor Brian Dolan for his unfailing help.

Introduction

First Principles

With the advent of the randomized clinical trial in the mid-twentieth century, the placebo effect mutated from a traditional resource of medicine into a confounding factor and, simultaneously, a potential object of theoretical interest. The transformation was marked by Henry Beecher's historic paper "The Powerful Placebo" (1955), with its finding that an average of 35% of the subjects in an array of studies responded to placebo as to an active medication, a figure that ignores alternative explanations for improvement such as the natural course of disease.[1] But while some have argued that the placebo effect dwindles almost to zero if correctly measured, over the decades since Beecher it has withstood all challenges, proving its ability to influence subjective experience and—more recently—to induce actual physiological changes.

Beecher's interest in placebos went back to his experience as an army doctor in World War II, when, lacking morphine to treat the wounded, he was reduced to injecting them with water. Exactly why this worked no one really knows, nor of course can the situation be replicated experimentally, but in all likelihood the strong expectation of relief and the evocative nature of the medical ritual itself contributed. Both expectation and "the power of context"[2] have since been identified as sources of the placebo effect, and both are of interest to this study.

As the RCT established itself as the definitive test of efficacy, medical research turned in that direction, and as the results stacked up, the placebo came to resemble a super-pharmaceutical capable of mimicking the effects, including even the side effects, of all kinds of drugs per se. Little wonder public fascination with the placebo effect centers on this virtuoso performer to the exclusion of background factors. But there is much more to the placebo effect than pills. If a placebo is inactive in itself, then its efficacy, if any, must arise from the context surrounding it, including in the first instance the medical encounter.

> How should an intervention (e.g. a saline injection) produce an effect if it is objectively without a specific effect? There now seems to exist a consensus among placebo researchers that what we call placebo effects is a heterogeneous class of psychobiological events attributable to the overall therapeutic context. The placebo intervention by itself should not produce any effect (otherwise it would not be a true placebo); it completes a complex therapeutic situation.[3]

Even the prescription of a pill takes place in a rich setting, a "complex therapeutic situation." But once we begin to see the placebo effect as in good part a social proceeding, we become aware that it does not stop at the doctor's door.

Our membership in a group can color our experience quite as much as a pill can; in fact, our notion of what a pill can do for us may well have been socially formed in the first place. We tend to expect the benefits others seem to experience. Recently the German Medical Association encouraged its own membership to use placebos more freely in clinical practice, and to recommend them to patients not as medications per se, which would be grossly deceptive, but as agents that have been shown to work with *other patients*.[4] (The same careful form of words, an equivocation that is also a direct appeal to our social nature, has been used in studies of the placebo effect.) This book looks into ways in which our affiliations with others support the placebo effect, or its nocebo equivalent, and extend its operation well beyond the clinic. Some might call the derivative experience of the placebo effect—as when a person reports benefits from an inert pill said to work with others— a secondary event, but I see no reason to privilege some other event as true and primary; it is not as if we were patients first and members of society second. Some speculate, indeed, that the placebo effect is social in origin, likely to have evolved "from social grooming in apes and altruistic behavior in early hominids. An individual who trusts a member of his own social group, whether a shaman or a modern doctor, has surely an advantage over those who lack this mental disposition."[5]

By consensus, it was the investigation of animal magnetism—a universal fluid postulated by the shaman-like medical doctor Franz Anton Mesmer (1734-1815)—that first pin-pointed the placebo effect. In keeping with the model of the placebo effect as a phenomenon activated by the context in which it occurs, I will argue that the felt power of the theorized fluid derived from the theatrical rituals of Mesmerism, but secondly from the power, the "magnetism," of the Mesmer movement.

In the tradition of the inquest into animal magnetism, investigators in recent decades have conducted intriguing experiments on the placebo effect, many of which concern pain. But because pain in the laboratory—pain studied and controlled—differs from pain of unknown origin capable of exciting anxiety and even terror, the results of such studies might not apply extramurally even if the trickery that is so much a part of placebo experimentation were admissible in medical practice.[6] Conversely, some forms of the placebo effect do not really lend themselves to simulation. Precisely by removing subjects from the social world at large and placing them in controlled conditions—often isolated from one another[7]—in the interest of methodological rigor, experiments shut off channels in which the placebo effect normally operates. This book views the placebo effect more broadly, in keeping with the importance of social context. In a neglected classic of the placebo literature it is documented that surgical patients whose windows looked onto "a small stand of deciduous trees" recovered more quickly than strictly comparable patients in the same hospital who looked onto a brick wall.[8] Let that more open view represent the view of the placebo effect itself taken here. Because my purpose is not to generate findings that can be translated into practice—though many of the findings of placebo research could never be translated into ethical practice either—I will consider some fictional portrayals of the social character of wellbeing and illness, even of dying. My assumption is that the subtlety and richness of literature more than make up for its deficiencies as data. The scientists who investigated the strange medical fashion of Mesmerism in pre-revolutionary France, thereby identifying what we now know as the placebo effect, concluded that its power source was none other than the human imagination. If the placebo effect plays in some way on the imagination, it seems fitting to consult works that know the realm of imagination from within—works *of* imagination—in a study of it.

I understand the placebo effect as an experienced benefit derived not from the actual composition of a medication or treatment but from its imputed or reputed properties, the rite of its administration, or "the power of context."

The Social Component of the Placebo Effect

If the placebo effect depends on belief, so great was belief in the Greek medical tradition in much of the Mediterranean world in medieval times that physicians who inherited and spoke for the tradition stood at the head of their local community and were sought after by kings and caliphs. "It was the general outlook of a highly bookish age with its deep veneration for scientific attainments which entrenched the position of the medical art in popular conscience."[9] In time, however, such bookishness attracted skepticism and ridicule, as the portrait of the Physician in the *Canterbury Tales* depicts the man's book-knowledge, or show of such, as a tool of a lucrative trade; or as the wise Lady Folly in Erasmus's *Praise of Folly* prefers uncertainty and bewilderment to the delusions of academic knowledge. To Lady Folly, the doctor who commands esteem by engaging the beliefs of the community is simply a charlatan.

> Among . . . many different disciplines, those are most highly prized which come closest to common sense, that is, popular folly. . . . "The doctor of medicine alone is worth all the others put together." And within this profession itself, the closer a man comes to an ignorant, arrogant, inconsiderate quack, the more highly he will be esteemed even by princes seated in lordly estate. For medicine, especially as it is now practiced by most doctors, is nothing but a branch of flattery, like rhetoric itself.[10]

But according to the compendious *Anatomy of Melancholy* by Shakespeare's contemporary Robert Burton, a physician cannot help his patients unless he inspires belief. "'Tis opinion alone . . . that makes or mars physicians, and he doth the best cures, according to Hippocrates, in whom most trust."[11] If the best physician is trusted by the most people—one thinks of Maimonides with his extensive practice, shining reputation, Hippocratic lineage, and presumably deliberate exploitation of the placebo effect[12]—even "opinion" in this case seems to refer to a belief generally, not just privately held.[13] It does not make things simpler that the word "opinion" in Burton's time had disturbing connotations of popular delusion.[14]

The physician who woos and wins the opinion of the community can be shown as a confidence artist, a master of benign deceptions, a good practitioner, or even all at once, as in this comment put to paper in the mid-eighteenth century:

The principal quality of a Physician, as well as of a Poet (for Apollo is the God of Physic and Poetry), is that of fine lying, or flattering the patient. . . . And it is doubtless as well for the Patient to be cured by the Working of his Imagination, or a Reliance upon the Promise of his Doctor, as by repeated Doses of Physic.[15]

Yet if most of our ailments pass, the lying physician may be speaking the truth despite himself, and if wellbeing has something to do with integration into the community, the physician may do some good by speaking the community's language.[16] The ambiguity of the figure of the socially adept healer—now a Maimonides, now a fine liar, now a quack—prefigures the ambiguity of the placebo effect itself, covering as it does a spectrum of responses ranging from actual bodily changes induced by paradoxically inactive agents to the dubious benefits of sham medications. In turn, the ambiguity of the placebo effect leads investigators to claim that "Positive emotions and ideas can help to heal the body through the powerful placebo effect," only to subjoin that "the actual effects of [optimism] are difficult to prove or disprove"[17]; or to label psychotherapy as "the quintessential placebo" only to deny that it is in fact nothing but a placebo.[18]

Placebos per se range from saline injections and sugar pills to unnecessary antibiotics, from active drugs prescribed at sub-active levels to drugs that outperform inert pills only trivially, as with antidepressants in most cases. If people are drawn to antidepressants because they seem to work, perhaps it is also true that they seem to work because so much belief, hope, scientific prestige, popular mythology, moral ardor, and financial capital have been invested in them that they became, in all, a kind of movement;[19] from 1988 to 2008 their use in the United States quadrupled.[20] It is known that the benefits of antidepressants are largely an artifact of the placebo effect. By general agreement, a primary mechanism of the placebo effect is expectation, as when we experience a certain benefit because we anticipate it. But we may anticipate it not because we ourselves have enjoyed it before, but because we learn, imagine, or assume that others have. In the pages to come I look from different angles at this predisposition to feel what others feel, or are reputed or believed to feel. Perhaps the first to take notice of it was Montaigne: "I would rather live among people who are healthy and cheerful: the sight of another man's suffering produces physical suffering in me, and my own sensitivity has often appropriated the feelings of a third party. A persistent cougher tickles my lungs and my throat."[21] But was this Montaigne's peculiarity?

Evidence that it was not comes from the observation of Robert Boyle—one of the founders of modern science—that a hysterical woman witnessing another suffering a fit was often "infected with the like strange discomposure."[22] Exactly the same observation was made over a century later by a doctor who collaborated in the first research in England on the placebo effect.[23] Yet the social component of our experience—our felt response to the experience of others, whether we witness it or learn of it—is largely overlooked in the now-voluminous placebo literature. A recent paper on "Patients' Direct Experiences as Central Elements of Placebo Analgesia" reviews studies where patients treated with a placebo are told of the relief others derived from it; that is, the paper unwittingly introduces reported experience into "direct" experience itself. (Indeed, in one of the reviewed studies the patients are actually told about the *practitioner's* experience treating their symptoms.) In each case the patients tend strongly to report the same benefit that was reported to them.[24] Similarly, it seems that many ask a doctor for antidepressants after learning of others' experience with them—so-called word of mouth endorsements.

While controlled experiments bear at best a rough resemblance to life at large, studies where subjects learn in one way or another of others' responses are closer to life than studies that seal subjects in an information vacuum. And much as the words of others enter into our own—for "in real life people talk most of all about what others talk about—they transmit, recall, weigh and pass judgment on other people's words, opinions, assertions, information"[25]—so too are the responses of others likely to tint ours. As it happens, research into the placebo effect in the late eighteenth century targeted two medical fashions powered by pamphlets, newspaper reports, word of mouth, public wonder—the flow of charged information. As I will argue, under the influence of these public sensations many reported bodily sensations exactly like those others seemed to have, even though both fashions turned out to be medically baseless. A few years later, in 1811, occurred the first recorded usage of the word "placebo" in the sense of a medicine prescribed to humor the patient. True to the social character of the placebo effect, however, patients may report (and conceivably experience) improvement in order to gratify the doctor,[26] just as the doctor may prescribe a placebo in order to appease the patient.

An oft-cited example of the placebo effect is the stimulation coffee-drinkers derive, or seem to derive or in any case report, from deceptively labeled decaffeinated coffee. No doubt the drinkers expect stimulation from caffeine, but where did the expectation come from? Maybe from their own experience, maybe not. Coffee-drinking, after all, is a social act surrounded with anecdote, such as tales of students who stay awake for nocturnal study sessions by charging with caffeine. "People hear about, observe, and experience coffee-drinking in specific contexts, embedded within sociocultural networks of meaning,"[27] so that I know of the effects of caffeine by reputation and rumor as well as first-hand experience. Socially speaking, it is not so mysterious that I should seem to experience the effects the world around me ascribes to caffeine, even though it is not present. In that coffee-houses where newspapers were read and opinions exchanged (and where doctors met)[28] were focal points of the public realm as it took shape in eighteenth-century Britain, coffee is in fact historically associated in a strong way with the circulation of information.[29] And hearsay or reputation itself can tinge our sensory experience, as when headache sufferers give high ratings to a placebo packaged as a famous brand of aspirin they have never taken.[30]

It sometimes happens that a drug's presumed effect runs contrary to its pharmacological one. In the case of alcohol, for example, "the pharmacological effect . . . is to decrease sexual arousal. However, consistent with common expectations, the belief that one has consumed alcohol results in *increased* sexual arousal to erotic stimuli."[31] From the folklore surrounding alcohol I derive the fanciful notion that it heightens sexual arousal. Steeped in this sort of common knowledge, I may come to experience an effect other alcohol-drinkers are presumed to enjoy, even though it is the presumption rather than the alcohol that drives it.

Or consider the finding that "people taking red or pink placebo pills tend to feel stimulated, and those taking blue pills tend to feel more sedated, regardless of active ingredients."[32] One doubts these reported experiences trace back in each and every case to the pill-taker's actual history with red and blue pills respectively. The fact is that in the world around us, these colors have a certain emotional valence—red connoting heat, passion, energy (as in a sports car's red line), blue associated with coolness and languor (as in the blues), melancholy, or even steadfastness ("true blue"). Moreover, the original experiment with pink and blue placebos was complicated with social factors that seem to have gone unnoticed in the literature. Though blue is no more strongly associated with languor than red with passion, the "blue" response was much stronger—possibly because responses were measured after the medical students who served as the study subjects sat through an hour-long lecture.[33] Not only can a lecture be a sedative in itself, but the drowsiness of some members of the audience can readily communicate itself to others.

GROUPS, FASHIONS, MOVEMENTS

Cases of sensations transfused from one person to another are well known to both medical and social science. In a public place—say a school or a train station—someone seems to scent a noxious gas and falls ill, whereupon bystanders fall ill in turn and the phenomenon cascades, with one person after another sickening as a result of exposure to mysterious vapors that turn out not to exist. But there is no law of nature or human nature dictating that only sensations of illness can pass in this way. Spectators of the physical "crises" induced by the charismatic Mesmer when he fixed subjects with his gaze might well have felt similar forces shooting through their bodies. Yet if we had to be on hand to witness the sensations of others in order for them to affect us, the social sources of the placebo effect would be more limited than in fact they are. For we also respond to reports about others, as indeed the stories that surrounded the Mesmer phenomenon must have added immeasurably to its mystique.

Only because the experience of our very bodies is subject to social influences can the placebo effect act in this way. Among the evidence that our sensations are so subject I would include notes left by the eighteenth-century physician Johann Storch, in the German town of Eisenach, concerning the ailments of his female patients. Speaking of a flux in their ears, of womb cramps in their mouth, the women described many bodily experiences all but incomprehensible to us but to them as intuitive as the social medium of language itself. They experienced their ailments alike just as they spoke about them in similar ways. To listen to the women is indeed to learn a new language:

> The complaint about an inner flux was one of the most frequent reasons why women turned to the doctor. . . . The flux is a strange thing. It described a host of things. "Flux" is the name for pains a woman felt inside from matter flowing in her body. The women also spoke of "flux" when something flowed from their bodies. The word "flux" combined a subjective experience with a complex meaning. The women suffered from an inner flux, but at the same time they were fearful that this flux inside them could be "struck in," be driven back, become stuck. They suffered from the flux and from the fear that it might disappear.[34]

Like flux itself, the women's ailments seem flowing and indeterminate, so much so that without settled ways of talking about them they might be unable to pin them down at all. Today we would be more inclined to speak of mood fluctuations than flux, although these too are highly ambiguous virtually by definition. The inherent ambiguity of our self-interpretations, to say nothing of mood changes, gives the placebo effect a field to operate in. If not for the equivocality of such events, the placebo effect would have much less to work with. In 1963 a meta-analysis of 67 drug studies tabulated such side-effects of placebo treatment as "depression of the central nervous system" (the most common of all), "heaviness of limbs," "mental confusion," and restless legs, events no less obscure, perhaps, than flux in the ears, but more credible to us because more familiar.[35]

The placebo effect may escape our awareness not because it is too remote but too near, as near as our own moods and pains. The report on the pink/blue pill experiment published in 1972 remarks in passing that it is no coincidence "that the most widely prescribed drugs are those used to treat mild anxiety and minor pain—conditions that either remit spontaneously or respond to reassurance." We are awash in the placebo effect, it seems. Not that nothing has changed over the intervening decades. These days drugs for depression have taken the place of those for anxiety on the sales list, but, perhaps unsurprisingly, these popular compounds exhibit a strong placebo component of their own. (As I will suggest, the very knowledge that one's drug has conquered the marketplace and is being taken by millions of others, as if one were part of a movement, may serve to boost the placebo effect.) Even though the two conditions, anxiety and depression, are associated with one another and difficult to distinguish clinically,[36] it seems one is in the ascendant or in medical fashion while the other is not. Not only has consumer favor shifted over time from anti-anxiety to anti-depression medications, both of them in large part placebos, but it was during these decades of pharmacological revolution that placebos acquired their now-celebrated ability to mimic drugs.[37] Highly responsive to its surroundings, the placebo effect simulated the action of the compounds that had become the darlings of public enthusiasm.

Some would restrict the locus of the placebo effect to dummy medications while others insist it is the art or rite of medical care itself. But a doctor may prescribe an irrelevant treatment with all due care. Ordering vitamin injections that had no particular medical value but nevertheless seemed to help his patients, one doctor would tell them, "I'm going to have you get some B-12 injections. They have helped many other patients, but I cannot explain to you why they work and I cannot promise you they will work. I can simply say that many patients tell me they feel better and stronger after such a course of therapy"[38]—thereby arousing an expectation that the benefits others enjoy, they will enjoy too. (That injections per se raise higher expectations among American than Europeans suggests that patients even respond to the needle as members of their society. For that matter, people tend to attribute their health problems to the same causes as the groups to which they belong.)[39] In saying that he can't make promises but that vitamin injections reportedly work for other patients, the doctor issues a disclaimer that nevertheless acts as a potent recommendation; the statement itself is an injection of hope. Similar formulas are used in placebo research when the experimenter does not want to lie but also does not quite want to reveal that the treatment in question is medically null, which would defeat the expectation of efficacy. Instances of this genre are cited in the pages to come.

In keeping with "the power of context" and the importance of ritual, we may benefit not only from taking the same medications or pseudo-medications as others, but following the same procedures. Evidence from several clinical trials suggests that subjects taking a placebo on schedule enjoy better outcomes, including significantly higher rates of survival, than the less adherent, even when a number of variables are controlled for.[40] Though the exact reasons for this provocative finding remain unclear, it seems the ritual of pill-taking—following the same procedure as others—does count for something. Similarly, the efficacy of Alcoholics Anonymous may flow less from the specific twelve principles to which members pledge themselves than from the communal nature of the pledge; by committing themselves identically in a solemn and ritualistic manner, members escape their own isolation and form a group that strengthens each and every one of them. As we will see, anonymity also rules over More's Utopia, a showcase of both good health and unanimous ritual.

"In the beginning was not the word but the group," suggests a searching analysis of the placebo effect.[41] In the course of a meditation on the same subject, a doctor notes that some of his patients in the 1960s refrained from alcohol for long periods when they joined not Alcoholics Anonymous but the Black Panthers.[42] So too, for veterans who have sustained psychological injuries in war mutual support may activate a potential for recovery—so-called "healing through community."[43] On the other hand, it remains unproven that support groups can improve survival rates for breast cancer patients; and while such groups may help men who choose to leave early-stage prostate cancer untreated,[44] the same men would most likely not have become cancer patients in the first place if they hadn't fallen in with the screening movement and sought out testing for a disease that can go untreated. Groups, then, take many forms, not all of them necessarily health-enhancing.[45] As social beings we are drawn to trends, fashions and movements with ambiguous potential, and are inclined to feel what others caught up in these forms of social life appear to.

According to some, underlying the placebo effect is the sense of being in the hands of a superior power, namely the doctor's.[46] While it is natural for doctors to regard themselves as indispensable to the placebo effect and central to the patient's experience, they do not actually have to be in the picture for someone to enjoy the placebo effect (as in the case of coffee with presumed caffeine), and at this moment few doctors are ready and willing to take charge of the patient as implied by the authoritarian model of superior power. The doctor who confesses, "I cannot explain to you why vitamin injections work and I cannot promise you they will work" has not struck a very authoritarian pose. But maybe I can also enjoy a sense of being in the hands of something greater than myself by committing myself to a movement— joining a multitude of others inspired by common aims or passions. (The members of Alcoholics Anonymous have not only joined a movement but, by their creed, have placed themselves in the hands of a higher power—a double source of morale that helps account for the success of the AA method.) Mesmerism, the craze that first inspired investigation of the placebo effect, was nothing less than a movement, and it didn't hurt that the man who gave it its name played the maestro and professed to be in touch with a mysterious elemental power. Some trace the practice of psychotherapy to Mesmer. In that spirit I will pursue an analogy between a current mode of psychotherapy and Mesmerism, but also make the more general case that the popularity of psychotherapy has much to do with its cultivation of the placebo effect, which is the other side of the argument that being carried along by a popular movement can fuel the placebo experience.

Though no longer in vogue, psychoanalysis provides the precedent for talking therapies that are. While the cures wrought by psychoanalysis were never confirmed experimentally[47] and remain open to doubt, the method was sustained by a potent narrative, according to which a patient held captive by unprocessed conflicts from childhood comes to recognize the source of his or her troubles and is thereby cathartically released from them. From where does this notion of catharsis—the discharge of energy that theoretically transforms us from prisoners of our past to free agents—draw its appeal? The fact is that healing has long been thought to operate by clearing harmful things from the body, whether by means of purging, bleeding, or some other method. The history of healing, it has been said, consists largely of "cathartic methods of treatment."[48] By directing the flow of animal magnetism and inducing cathartic "crises" of his own, Mesmer purported to clear blockages from the patient's system without recourse to traditional medical methods. Like Mesmer, Freud adapted a principle that seems to have recommended itself to human intuition for as long as something like medicine existed. The authors just cited also argue, with evidence, that "until recently the history of medical treatments was essentially the history of the placebo effect."[49]

Rich Representation

Over recent decades many have sought, with reason, to rescue the placebo effect from the cynicism that once surrounded its use as a ploy to placate and deceive gullible patients. But I know of no more robustly straightforward defense of the value of pleasing the patient—and "placebo" means "I shall please"—than Rabelais'.

In contrast to those who portray physicians as flatterers and confidence artists, Rabelais, a physician himself, would have them serve patients in all sincerity by lifting their spirits, just as his own writings are intended to relieve depression and minister to human cheerfulness. Given the literal meaning of "placebo," Rabelais' position that the doctor should above all please the patient constitutes a warrant of the placebo effect.

> A physician, dressed up with the right mien and attire . . . could reply to those who found his role-playing odd: "I have put on such accoutrements not to show off and be pompous, but to please the patient on whom I am making a call, whom alone I seek entirely to please, avoiding all offense and irritation."[50]

The proper use of the placebo effect is not to exploit the patient by inducing belief in sham remedies but to encourage by every fair means. Rabelais makes the transacted nature of the pleasing effect quite clear, whether "such cheering-up results from the perceptions of the patient as he contemplates those qualities in his doctor . . . or whether it results rather from the pouring of the doctor's spirits . . . into the person of his patient."[51] Either way, each feels what the other does.

Our liability to deception was undoubtedly known to Rabelais, as the human appetite for delusion constitutes a commonplace of the satiric tradition and indeed literature in general. Literature, however, laughs and weeps at human suggestibility as medical research will not permit itself to do. It laughs, as in a *Decameron* tale where a group of rogues set up the foolish Calandrino by accosting him one by one and asking him if he is all right; by the third cue that something is wrong, he "was quite certain he was ill."[52] A doctor, the rogues' confederate, then tells him the reason he feels so sick: he is pregnant. Literature also weeps, as when Eve accepts the apple in *Paradise Lost* (the subject of chapter 4). Satan doesn't just extol the supposedly magic fruit but cites his own experience of its uplifting effects; Eve's momentary sense of uplift upon eating it, obviously modeled on the serpent's report, is the placebo effect without the name. If we can be led to confuse alcohol for a sexual stimulant, perhaps "our general mother" could confuse a common apple for a psychotropic one.

I look to literature in these pages, then, because it knew of our propensity for delusion as well as the social character of our experience well before these matters came before the bar of science and were isolated and verified experimentally. But there is another reason.

Two decades ago a paper in *Science* memorably demonstrated a connection between social bonds and health, such that the less socially integrated are even more likely to die: a striking illustration of the medical import of social forces.[53] But how is the nature of a social bond to be assessed? Here the paper is at its weakest, it seems to me. Time and again it refers to the "quality of social relationships" as if that elusive something were as plain as a box on a questionnaire. The information available in forms and surveys about the quality of a relationship—its dynamics and differentiae, its intricacies—is itself of poor quality. A significant contributor to inflated estimates of the prevalence of depression in the United States is the crudity of instruments used to measure it.

So argues a book published a few years ago that takes on the diagnostic system of the authoritative *Diagnostic and Statistical Manual*, contending that a checklist of symptoms for such an ambiguous condition as depression is too schematic and leads to false conclusions. As the authors make clear, the diagnostic question is not which presumptive symptoms of depression exist, but whether the symptoms are in fact indicators of excessive, chronic or uncaused sadness, a determination that can be made only if the history behind the symptoms is taken into account. By way of illustration they offer some hypothetical cases, one involving the collapse of "a passionate romantic relationship," another the "loss of a valued job," another the receiving of a "life-threatening medical diagnosis" by a loved one, each of which might incite a reaction of profound sadness.[54] But case histories sketched in two or three paragraphs are themselves bone-thin compared to the richness of literary representation. In the literature of the imagination we will find a feeling for detail, for the specificity of cases, and for ambiguity, each of which tends to be missing from abbreviated reports, and all the more from the statistical language in which medical findings are now so often cast.

The most pointed discussion I have encountered of the biomedical influence of social forces occurs, in fact, in a novel. When the aging protagonist of Wallace Stegner's *The Spectator Bird*—a man of bitter meditations, graveyard humor, and Danish ancestry—receives a questionnaire in the mail asking about his self-esteem on the theory "that a decline in self-esteem is responsible for many of the overt symptoms of aging," it sends him into a rage:

> I looked at the questions and threw the thing in the fireplace. Another of those socio-psycho-physiological studies suitable for computerizing conclusions already known to anyone over fifty. Who was ever in any doubt that the self-esteem of the elderly declines in this society which indicates in every possible way that it does not value the old in the slightest, finds them an expense and an embarrassment, laughs at their experience . . .? The poor old senior citizen has two choices, assuming he is well enough to have any choices at all. He can retire from that hostile culture to the shore of some shuffleboard court in a balmy climate, or he can shrink in his self-esteem and gradually become the cipher he is constantly reminded he is.[55]

In time, socio-psycho-physiological studies became the currency of research into the placebo effect, and because I cite many such, I think it best to temper their abstraction with particularity. Before looking into the social character of the placebo effect as it was identified in the late-18th century and then tracking some of its manifestations today, I will therefore examine a few socio-medical transactions in literary works of universal renown. In Book Four of the Odyssey the action of a certain benign Egyptian drug seems scarcely distinguishable from the ritual of its consumption. In More's remarkable Utopia, the action of an also-benign suicide drug seems similarly supported by ritual. In the twisted world of *Hamlet*, however, it is the poisoner of Hamlet's father who invokes the social nature of wellbeing, espousing conventional social remedies for the young man's melancholy—behavioral antidepressants, we might call them.

Medicine and literature are kindred arts. Apollo presides over both Physic and Poetry, after all.

Chapter One

Heartsease
Medicine and Social Context

> One of the foundational works of our literature shows, though it does not
> state, that a medication's mode of action may be social.

Among the earliest allusions in Western culture to the art of medicine is a passage in the Odyssey referring to a certain mysterious drug acquired by Helen in Egypt and now administered both to her husband Menelaus and his visitors in their wine:

> Into the wine of which they were drinking she cast a medicine
> of heartsease, free of gall, to make one forget all sorrows,
> and whoever had drunk it down once it had been mixed in the wine bowl,
> for the day he drank it would have no tear roll down his face,
> not if his mother died and his father died, not if men
> murdered a brother or a beloved son in his presence
> with the bronze, and he with his own eyes saw it. Such were
> the subtle medicines Zeus' daughter [Helen] had in her possessions,
> good things, and given to her by the wife of Thon, Polydamna
> of Egypt, where the fertile earth produces the greatest number
> of medicines, many good in mixture, many malignant,
> and every man is a doctor there and more understanding
> than men elsewhere.[1]

In the tradition of Homer, many Greeks were to place the origin of medicine and pharmacology in Egypt, among them Herodotus, whose comment on the Egyptians' medical knowledge may have been inspired by this fabulous passage.[2]

While reminding us of the magical herb that protects Odysseus against enchantment at the hands of Circe[3] or the lotus that enchants men's minds by destroying their desire for home, Helen's exotic drug nevertheless has a story of its own, and so too is its use embedded in a rich narrative context.

Prompted by Athena to seek for news of his father, Telemachus journeys first to Nestor in Pylos, then in company with Nestor's son Pisistratus to "the palace of the king whom Zeus loved" (4.44), Menelaus, in Lakedaimon. There the young man is received royally, as befits the son of a king; indeed, he is recognized as Odysseus' son even without announcing himself, which ought to lay to rest his doubts about the identity of his father (1.215-16). Menelaus professes a strong and abiding love of Odysseus and laments his disappearance, stirring up in the others a desire for weeping; at which point Pisistratus too weeps for his brother Antilochus, lost in the Trojan War. Once the tears flow, Menelaus proclaims it is time for the washing of hands, and then dinner, whereupon an attendant "poured water for them to wash with. / They put their hands to the good things that lay ready before them" (4.217-18). It is at this point, then, as a refinement of the dining ritual, that Helen introduces a subtle narcotic into the wine. The action both before and after speaks to the dependence of the medical art on the social context in which it is practiced, a point under discussion even now.

While Helen's drug is described as extraordinary, the occasion framing its use is extraordinary in its own right, if differently so. Never before have these four persons stood in one another's presence, nor perhaps will they ever again, after this one visit. The moment is thus lifted out of the ordinary flow of events into the light of the remarkable. The presence of these people in the same place at the same time is itself a potent mixture. We note, for example, that even before the men drink the heartsease they have eased their hearts by pouring out their grief. In other ways, too, the action of the drug is supported by the setting in which it is taken, its potency activated by an elaborate and highly evocative context of ceremony. After all, there could be no better place for easing the heart than the palace of "the king whom Zeus loved," a palace almost Olympian in its amenities (and the Homeric gods indeed live at their ease);[4] but it is not the opulence of the palace alone so much as the decorum of its rituals that seems to underwrite the power of heartsease. If the drug is not administered covertly but given and received like a gift, the acts of giving and receiving are ritualized in the Odyssey as if they were the crux of social life. Helen, the giver of the drug, received it herself in Egypt, just as Menelaus received the mixing bowl fashioned by the gods' artificer that he gives Telemachus in turn (4.615-17). It is the most splendid gift in his possession.

"Now when she [Helen] had put the medicine in, and told them to pour it, / taking up the story again she began to speak to them" (4.233-34). As the use of heartsease is folded into the ceremony of drinking, so it is mixed into

the wine openly, decorously, not in secret. It is not enough for Helen to intro-
duce the drug into the wine, but she must be seen to do so, her action in this
respect standing in direct contrast to Circe's surreptitious admixture of drugs
into the wine she serves Odysseus' men (10.236) to make them forget their
homeland.[5] Perhaps if Helen's medicine had not been so fully integrated into
the rituals of decorum and hospitality, it too would have produced only some
form of amnesia rather than the profound pleasure in hearing stories that
seems to be its actual effect. Indeed, immediately after serving the enhanced
wine, Helen launches into tales of Odysseus.

> Son of Atreus, dear to Zeus, Menelaus, and you who
> are here, children of noble fathers; yet divine Zeus sometimes
> gives out good, or sometimes evil; he can do anything.
> Sit here now in the palace and take your dinner and listen
> to me and be entertained. What I will tell you is plausible.
> I could not tell you all the number nor could I name them,
> all that make up the exploits of enduring Odysseus . . . (4.235-41)

If eating and drinking, not in the manner of the riotous suitors but as properly performed, have a ritual character, the telling of stories completes the
occasion.[6] It is after eating and drinking that Odysseus himself, in the land of
the Phaeacians, asks the blind bard to sing the story of the Trojan Horse, the
same ruse of war Menelaus tells of in Book Four.

If indeed "all sorrows can be borne if you put them into a story or tell a
story about them" (as Hannah Arendt has written, attributing the saying to
Isak Dinesen),[7] then the anodyne effect of heartsease in fact resembles that of
narrative—resembles it so suggestively as to make the two virtually one. Stories, on this showing, reconcile us to things as they are, in particular to grief
and loss. The Trojan War itself became a great source of grief and loss even
for the victors, though by the same token those sufferings can be put into the
form of stories. What makes Odysseus' disappearance all but unbearable for
his loved ones is that precisely because his fate is unknown his story cannot
be told. As we would put it today, they lack closure. If Odysseus had died in
Troy (says Telemachus in Book One), his fame would have survived, which
is to say that stories would have been told of him. "But now ingloriously the
stormwinds have caught and carried him / away, out of sight, out of knowledge, and he left pain and lamentation to me" (1.241-43). While death leaves
pain and grief, they are alleviated by story; but of the fate of Odysseus no
story can be told.

Neither Menelaus nor Telemachus nor Pisistratus has a brother killed in front of his eyes while under the influence of heartsease. Indeed, nothing but storytelling takes place when the drug comes on, and storytelling to the ear very much like all the other narratives in the Odyssey. Menelaus himself, though drugged, tells the tale of Helen circling the Trojan Horse, calling out to the warriors inside in the voice of their wives like some ancient Tokyo Rose—tells of this event, the very crisis-point of the Trojan War, the moment on which its outcome depends, exactly as if he had never drunk heartsease at all. While we do not discover whether heartsease can really make a man indifferent to the murder of a brother, we have already learned that story can reconcile a man to such a loss, or at least begin to.[8] The murder of Menelaus' brother Agamemnon is probably the most frequently told story in the Odyssey; and while Menelaus laments his death, he does not (as he tells us) grieve it so much as the disappearance of Odysseus, who, precisely because he *has* disappeared from view, cannot be commemorated in story (4.104-10). Story itself is consolation, is heart's ease. Pisistratus laments the loss of his brother Antilochus, "surpassingly swift of foot, and a fighter" (4.202), though his grief is made more bearable by being cast into words, and within a few lines he is eating dinner.

By virtue of the heartsease ritual, then, three men with their solitary sources of lamentation—Telemachus, consumed with grief and self-doubt; Pisistratus, "thinking in his heart of stately Antilochos" (4.187), whom he never saw as Telemachus has never seen Odysseus; Menelaus, mourning all who died in his cause (4.97-102)—put aside their private thoughts like members of a single company. In this special setting, to feel what others feel means to be released for the time being from the imprisonment of private sorrows.

In Plato's *Laws* those afflicted with the impulse to rob temples are advised to perform cleansing rites and seek the company of men with a reputation for virtue; if they are lucky their "disease" will abate.[9] It seems the right sort of company and the right sort of rituals can have a curative effect. In the heartsease episode of the Odyssey a similar point is made dramatically. However potent the drug is said to be, its effect of releasing from sorrow appears to derive from the ritual of its use, and suggestively resembles the effect of the exemplary communal activity of storytelling. Not until the following day—after the drug has worn off—does Telemachus hear the account of Menelaus' journey home to Sparta, in the course of which the Old Man of the Sea reports that Odysseus is being held captive by Calypso. After attending to the tale and receiving some of the information he came for, Telemachus begs leave to depart, first telling his host, "I could well be satisfied to sit here beside you / for a year's time, without any longing for home or parents, such strange pleasure do I take listening to your stories / and sayings" (4.595-98). Even allowing for an element of courteous hyperbole, it seems a well-delivered tale has a pleasure-giving effect in and of itself that is much like the described effect of an exotic medicine. The power of the preparation derives less from the soil of Egypt than from the ceremonies surrounding it.[10]

Only a few lines into the Odyssey it is reported that Calypso "works to / Charm [Odysseus] to forget Ithaka" (1.56-57), but without success. We note that Calypso uses no drug at all, merely her own attractions. Conversely, a scene in the Iliad shows the sheer act of conversation as so pleasurable and soothing that it brings enjoyment to a man shot with a barbed arrow. The wounded Machaon (son of the legendary healer Asclepius, as it happens) is borne back to Nestor's shelter, where a slave duly prepares a potion of wine with white barley and grated cheese admixed. After the two men slake their thirst they "began to take pleasure in conversation, talking with each other" (11.642), meaning that Machaon is relieved by speech alone, with no need of the drug given to another arrow-wounded warrior later in the same book.[11] If conversation in and of itself can assuage the pain of a barbed arrow, it is surely possible for story to dispel grief.[12] "Words can be powerful placebos."[13]

⊟

In the world of the Socratic dialogue we learn of another medication requiring a context of ritual to take effect. In a ruse to bare the soul of Charmides (who seems to be about the age of Telemachus), Socrates pretends to know a certain incantation that activates a certain drug that relieves headache:

> And I said [the drug] was a certain leaf, but that there was a certain incantation in addition to the drug, and that if one chanted it at the same time that he used it, the drug would make him altogether healthy, but without the incantation there would be no benefit from the leaf.[14]

Yet the dialogue never gets around to the incantation or the leaf—or the headache, for that matter—quite as if these were pretenses to set up the matter it is interested in: health, or soundness, of soul, for "everything starts with the soul." The actual existence of both drug and incantation is left in question. (Interestingly, the dialogue also refers at a number of points to the false practice of medicine.) In Book Four of the Odyssey we are shown both the Egyptian drug and the ritual context in which it is taken. In this case, story as opposed to incantation provides the enchantment that gives the drug effect; while decorum requires that Menelaus not ask his guest the reason for his visit until the next day, story speeds the flow of time and imposes its wonder on the listener's troubled soul. The tales of Odysseus' exploits behind enemy lines and inside the Trojan horse—both illustrating his aptitude at unconventional warfare—shine all the more brilliantly in that they stand out against the mystery of his disappearance.

Today we might say that heartsease acts through the mechanism of the placebo effect—that its power lies not in its specific composition so much as in the expectations raised by and the ceremony surrounding its use. (Given to Helen in the first place and mixed in a bowl like the one given to Telemachus upon his departure, the drug is set in a rich context of gift-giving; in a study of the placebo effect, receiving a medication from a doctor is likened to receiving a gift.)[15] If Helen's audience had not been primed for pleasant effects by the beauty of the palace and its furnishings and rituals, and by seeing her lace the wine, it might not have experienced quite those effects. Expectation has been identified as "the main factor in placebo responsiveness,"[16] which is not to say that the placebo effect necessarily reduces to a kind of bait-and-switch experience—being led to anticipate one treatment only to receive another. It is to say that the placebo effect has social sources and draws on the ambiguous potential of our nature as social beings. Perhaps the mysterious reputation of Egyptian medicine as cited by Homer traces not only to the composition of Egyptian drugs per se but the uniquely collective nature of medical knowledge in a realm where "every man is a doctor," as the uniquely good health of More's Utopia (see next chapter) reflects the unanimity of the citizenry.

If a Telemachus desperately uncertain of himself finds solace and assurance of his identity in the company of others,[17] social ties may be health-giving in and of themselves, as Durkheim showed in his great study of suicide a century ago—although the effect of companionship on our wellbeing goes beyond even this, as both literature and medical literature suggest. It is now recognized, by many at least, that the way in which medical care is administered can be as important as the treatments themselves, just as the rich social ceremonies surrounding the use of heartsease, and the storytelling it is so closely associated with, signify as much as the drug. Indeed, Menelaus displays to the troubled Telemachus many of the qualities that patients now look for in their doctor, in particular warmth, concern, command of information (in this case the information that Odysseus is marooned on Calypso's island) and the ability to relate it, as well as an unhurried manner,[18] to which we can add honesty. "Tell me the whole truth," says Telemachus (4.351).

A recent paper in the *Journal of the Royal Society of Medicine* argues that the very setting and manner of healing may contribute to therapeutic efficacy.

> That aspect of healing that is produced, activated, or enhanced by the context of the clinical encounter, as distinct from the specific efficacy of treatment interventions, is contextual healing. Factors that play a role in contextual healing include the environment of the clinical setting, cognitive and affective communication of clinicians, and the ritual of administering treatment. Contextual healing is precisely what has been off the radar screen of scientific medicine.[19]

While the claim that the placebo effect can actually heal, as opposed to relieving symptoms, remains open to question, the importance of context in generating and supporting the effect is undeniable. In Book Four of the Odyssey, accordingly, hospitality itself, with its rituals and affective communications, serves a sort of placebo function. We might even consider the palace of Menelaus, with its high roof, gleaming interior, excellent baths, staff of attendants, and general good order, as a variant hospital in which a heartsick Telemachus finds strength, if only by being recognized and treated as the son of Odysseus and thereby coming to know himself as such. While Homer foregrounds and even romanticizes the Egyptian anodyne—the treatment intervention—and leaves its dependence on social ceremonies a matter of inference, some awareness of "the power of context" seems inscribed into the Odyssey.[20]

Chapter Two

Suicide in Utopia

As with heartsease, the ritual surrounding the Utopian suicide drug is itself medicinal.

The *Science* paper documenting a link between social integration and health ends by noting a fracture of social patterns in the very time and place of its writing, such that "just as we discover the importance of social relationships for health, and see an increasing need for them, their prevalence and availability may be declining."[1] In Thomas More's Utopia, not only do social integration and health go hand in hand, but the islanders' way of life is spared this kind of corrosive attrition. Their arrangements are good, and being so, remain unchanged over the generations.

A pageant of sameness and harmony, Utopia is "like a single family,"[2] but one free of the conflicts and tensions of familial life as we know it. The collectivism of Utopian life is a legacy of Plato, whose ideal commonwealth, Magnesia, is constructed on that very principle. "In short," says Plato, "we must condition ourselves to an instinctive rejection of the very notion of doing anything without our companions; we must live a life in which we never do anything, if possible, except by combined and united action as members of a group."[3] The unanimity of Magnesian life, expressed in music, dance, and civic ritual, seems conducive to collective health, Plato's intent in designing the commonwealth being the quasi-medical one of preventing some evils and curing others. In Utopia, for its part, all adhere to the same healthy way of life, or it may be that their way of life is healthy precisely because they hold to it in common—that is, because their unanimity exempts them from the greed and envy that are the ruin of Europe. The Utopians prize a good that cannot be taken from one person by another to be added to the latter's hoard: simple wellbeing. The principal narrator of *Utopia* and our source of information about the happy society is a traveler named for the archangel Raphael, by tradition a healer.[4]

Unlike those for whom health is the absence of pain or disease, the Utopians deem health a positive state and relish it for its own sake. "Health itself, when not oppressed by pain, gives pleasure, without any external excitement at all. Even though it appeals less directly to the senses than the gross gratifications of eating and drinking, many consider this to be the greatest pleasure of all. Most of the Utopians regard this as the foundation of all the other pleasures" (p. 74). Arguably, the Utopians rate health so highly because the pursuit of so modest, moderate and innocent a good offends no one. The cultivation of health is socially ideal and therefore celebrated, which would explain how it is that the Utopians are so conscious of just what anyone else might take for granted—the quiet wellbeing of their own bodies.[5] "The idea that health cannot be felt they consider completely wrong" (p. 75). In finding enjoyment in health, they feel what they are told and primed by their way of life to feel and what others around them also seem to feel. A socially fueled placebo effect underwrites their experience of their physical selves.

And if and when the time comes for the Utopians to end their lives, they do so just as socially as they lived.

⊟

Among the revolutionary features of Utopia, along with the abolition of private property, money, and monarchy, is legal suicide. But revolution here means a return to ancient models and precedents, for just as More owed the literary genre of the ideal commonwealth to Plato, he revived the principle of licit suicide in special cases codified in Plato's *Laws*.[6] Not that licit suicide had no reality outside the pages of Plato. In Athens and certain Greek colonies the authorities maintained a supply of hemlock for would-be suicides, requiring only that they make their case to the Senate and win its permission to end their lives.

> Whoever no longer wishes to live shall state his reasons to the Senate, and after having received permission shall abandon life. If your existence is hateful to you, die; if you are overwhelmed by fate, drink the hemlock. If you are bowed with grief, abandon life. Let the unhappy man recount his misfortune, let the magistrate supply him with the remedy, and his wretchedness will come to an end.[7]

Utopian policy toward suicide is generally similar, although in this case the sufferer does not petition the state but the state the sufferer, which takes the initiative out of private hands and makes the process less whimsical. Indeed, the procedure surrounding the use of the lethal drug is as medicinal as the drug itself.

In Utopia the prohibition of unauthorized suicide is not just implied but emphatic. The Utopians are not allowed to make an end of themselves at their own discretion. Those suffering the pangs of despised love or any of the other causes of despair named by Hamlet in his meditation on suicide, those whose existence has simply become hateful, are ineligible for suicide. But precisely because the practice is strictly regulated, those who do end their existence with the blessing of the Utopian state must find great comfort in that blessing, even as they find oblivion in the drug dispensed by the attending priests and civic officers. The commonwealth's approval is itself an anodyne, all the more because (unlike the vote of a Senate), it is conveyed intimately, in a ritual that serves as a preparation for death. Discussed as a special case of care for the sick, the practice of authorized suicide in *Utopia* suggestively illustrates the social component of medicine, perhaps even implying that the power of the drug in question reflects the power of the figures who administer it.

Only the incurably ill in intractable pain are allowed to end their lives in Utopia, and to make the decision easier, officers of the commonwealth visit them at their bedside, place a mysterious narcotic at their disposal, and urge them with powerful arguments to use it.

> They remind him that he is now unequal to any of life's duties, a burden to himself and others; he has really outlived his own death. They tell him he should not let the disease prey on him any longer, but now that life is simply torture and the world a mere prison cell, he should not hesitate to free himself, or let others free him, from the rack of living. This would be a wise act, they say, since for him death puts an end, not to pleasure, but to agony. In addition, he would be obeying the advice of priests, who are interpreters of God's will; thus it will be a pious and holy act.

> Those who have been persuaded by these arguments either starve themselves to death or take a drug which frees them from life without any sensation of dying. But they never force this step on a man against his will. . . . The man who yields to their arguments, they think, dies an honourable death; but the suicide, who takes his own life without approval of priests and senate, him they consider unworthy of either earth or fire, and they throw his body, unburied and disgraced, into the nearest bog. (p. 81)

The reprobation of willful suicide in Utopia measures the approval of suicide in cases of incurable bodily suffering. The delegation of visitors to the incurably sick includes the most revered of all figures in Utopia—priests—and the arguments deployed have the force of first principles. Though to us this procedure may well seem heavy-handed, the visitors' intent is not simply to persuade sufferers to end their life but to assure them of their society's regard if they do. The illegal suicide is tossed into a bog. The legal suicide dies with full honors in the arms of the commonwealth itself, and the philosophical medicine administered beforehand constitutes an essential preparation for the medicine per se. If the sufferer dies painlessly, one reason may be that he is assured that suicide under these circumstances is perfectly permissible and he has nothing to fear beyond the grave. The giving of these assurances may be as important as the composition of the drug itself. The show of concern by honored citizens and the release from fear of punishment and opprobrium, together with the expectation of swift death, ease the passage from life.

The Utopians' approval of suicide in the case of terminal illness accords with the value they attach to pleasures of the body, the foremost of which, as we know, they judge to be health. "They nearly all agree that health is crucial to pleasure. Since pain is inherent in disease, they argue, and pain is the bitter enemy of pleasure just as disease is the enemy of health, then pleasure must be inherent in quiet good health" (pp. 74-75). Thus the rational arguments recited at the sufferer's bedside grow out of the principles of Utopian life, and precisely *as* arguments are much in the style of this philosophical people. Of Utopia it is said that "there's hardly a country in the world that needs doctors less" (p. 79), and the source of the islanders' good health, a reader concludes, is not simply their moderation or dietary practices but their way of life itself, a regimen in theoretical accordance with Nature, and to which all adhere. The Utopians live as one. Significantly, in More's text no citizens have names or any other mark of individuality. Like the officers of persuasion who visit those suffering incurably, the inhabitants of Utopia are described in common, as "they." "They were delighted" to discover Hippocrates and Galen (p. 79). The suicide who dies with state approval remains part of this social body, and without such moral comfort he or she might be in no position to receive the comfort of the suicide drug.

With few laws bearing down on them, the Utopians behave well spontaneously, but the reason their society can trust them to do so is that the same wholesome principles are drilled into everyone from an early age, making laws largely unnecessary. A perfect image of Utopian spontaneity appears in a description of the islanders' way of worship. "As the priest in his robes appears from the vestry, the people all fall to the ground in reverence" (p. 105). Thus, when priests visit the incurably sick, they have been habituated by civic training to submit all but automatically to these figures; or more favorably put, they are in the best possible position to accept the assurances that ending their life is not only a permissible and rational but a pious act. As the populace unanimously falls profoundly still before the priest in church, so the dying fall still once and for all after accepting both the arguments and the drug dispensed by a priest. Moreover, considering that the priests are also responsible for the education of the young in Utopia, the cleric who presents a set of arguments to the dying might as well as be the instructor who taught them to think in the first place. The suicide drug is embedded as deeply in the context of Utopian life as the poisoned wine in the text of *Hamlet*.

With its citizens subjected to an identical conditioning regimen from their early years and as a result so indistinguishable that they can be referred to generically, Utopia is a model of social integration. Its citizens, far from suffering the lawlessness detailed in Book One or the loss of belief in law itself later to be known as anomie, have the unwritten laws of their society engraved on their hearts. The willful suicide who defies these laws is cast out in death more or less as bad humors or harmful excesses were expelled from the body in accordance with medical theory grounded in Hippocrates and Galen, whose writings we are told the Utopians received with delight. It follows that only the most desperate or rebellious Utopians would commit suicide without authorization, knowing the disgrace and fearing the damnation that awaited them.

But by the same token, when the priests approve the act of suicide, the very majesty surrounding these figures assures the sufferer that the society that has been his support in life will continue to esteem him in death. Once his mind is at ease, he is ready to take the poison that frees him from life "without any sensation of dying." The drinking of this mysterious draft is the consummation of a process that constitutes a last rite, a sequence as powerful in its own way as the church services over which the priests preside, and which are the cardinal ritual of Utopian life. Toward the end of these services the Utopians pray to be received by God "after an easy death" (p. 106). For those in suffering, the administration of poison provides just that—by comforting the mind as well as the body. The drug comes as a blessing, not least because its use actually has been blessed. As if in confirmation of the principle that death itself has a social component, euthanasia, a happy death, has come to a citizen of Utopia, a happy place.

☒

I have suggested that the power of the heartsease, in Book Four of the Odyssey, is tantamount to the ritual surrounding it. After the drug is introduced into the wine, we hear nothing more of it. What we hear are stories—stories embedded in the rich ceremonies of hospitality and producing, in and of themselves, something very like the magical consolation said to be the drug's effect. The drug's mechanism would appear to be social.

So it is, more or less, with the Utopian suicide drug. About both preparations—one lethal, one said to produce forgetfulness, Lethe-like—we know very little, and in both cases it is the context of the drug's use that bears the brunt of the description. Consider the power of the ritual surrounding the administration of the suicide drug, a ritual presided over by the same priests who see to the instruction of the young (the process that makes legislation all but unnecessary) and conduct the religious services at the heart of Utopian civic life. "The priests are of great holiness," writes More, "and therefore very few" (p. 101). The exalted stature of the priests would make their visit to the incurably sick and their ritual recitation of the reasons to die, culminating in the offer of the lethal cup, all the more compelling. If patients who feel they are in the hands of a superior power (the doctor) are well positioned to enjoy the placebo effect, the dying who accept the cup from the priest commit themselves to a superior power indeed. The drinking of the poisoned draft represents the last step of a process in which the pain of leaving life is taken away. In a sense, indeed, the citizens of Utopia have been in training all their lives for the proffered cup. From childhood they are taught the proper valuation of life, such that "they don't hold life so cheap that they throw it away recklessly, nor so dear that they grasp it greedily at the price of shame when duty bids them give it up" (p. 93), a philosophy that comes to their aid when they find themselves "unequal to any of life's duties, a burden to [self] and others" (p. 81). The incurably sick are thus superbly prepared to release their grasp on life. If heartsease is embedded in ritual, the Utopian suicide drug is embedded in the islanders' way of life itself.

In some literary cases circumstance doesn't support the effect of a drug or medicine so much as it acts like a medicine in its own right. Iago poisons Othello's mind without the use of any foreign substance; after the General falls to the ground in a fit as if in reaction to some powerful drug, Iago exclaims, "Work on, / My medicine, work!" The only medicine at work is the power of his obscene suggestions. We turn now to *Hamlet,* where, by contrast, one familiar with the use of actual poison prescribes purely social measures in an attempt to allay the hero's melancholy.

Chapter Three

Malady and Remedy in *Hamlet*

Hamlet's melancholy resists social remedies.

More than two millennia after the *Odyssey*, heartsease reappeared in an extraordinary compendium of medical thinking that jointly considers mind and body, as well as person and social circumstance: Burton's *Anatomy of Melancholy* (first edition, 1621). Given Burton's interest not only in the divisions, manifestations and causes of melancholy but its remedies, it seems fitting that he should refer to an episode of the Odyssey that has Menelaus and his visitors going from grief to the pleasure of eating, drinking, and storytelling in a matter of a few lines. ("Surfeit of gloomy lamentation comes quickly," says Menelaus [4.103].) Indeed, as Burton sees things, heartsease is simply a poetic figure for the effect of convivial activities.

> *Jucunda confabulatio, sales, joci,* pleasant discourses, jests, conceits, merry tales, *melliti verborum globuli* ["dainty-fine honey-pellets of words"], as Petronius, Pliny, Spondanus, Cælius, and many good authors plead, are that sole nepenthes of Homer, Helena's bowl, Venus's girdle, so renowned of old to expel grief and care, to cause mirth and gladness of heart, if they be rightly understood.[1]

"Nepenthes" refers to Helen's heart-easing drug, evidently the same compound for which "Helena's bowl" stands by association—Burton's point being that it has no particular medicinal property or magic, that the remedy for grief is to be found in human company itself. Misery loves company, not necessarily spitefully. Perhaps, indeed, Menelaus, Telemachus, and Pisistratus are released from their sorrow in the sense of being able to feel the sorrows of others. Having never seen his father, Telemachus may better comprehend Pisistratus' grief for a brother he too never saw. Having had his life saved by Odysseus inside the Trojan Horse (4.280-84), Menelaus may better comprehend Telemachus' sense of loss without him. The company of others takes us out of ourselves. Avoid excessive solitude, says Burton, and "as much as in thee lies live at heart's ease" (II.124).

Burton's proviso "if they be rightly understood" evidently means that the Egyptian drug and associated lore need to be interpreted allegorically (and thus brought into line with Renaissance understanding); but except for recommending a kind of merriment not to be found in Book Four of the Odyssey,[2] his argument that the remedy for sorrow lies in social activities, not in mysterious compounds like heartsease, has considerable support in the text of the Odyssey itself. When Burton recommends as a specific for melancholy "a cup of good drink now and then" (2.120), he doesn't mean drink laced with a special anodyne, but even if the wine consumed by Telemachus, Pisistratus, and Menelaus—each of them with haunted with grief—had contained no such medicine, they would have been comforted with the ritual of drinking, as they flourish in each other's company and take pleasure and solace in the telling and hearing of stories whether under the influence of the medicine or not.

The convivial act of eating may also relieve grief—so suggests the famous or infamous tale of the widow of Ephesus told by Burton's Petronius. Such is the widow's grief that she fasts in her husband's tomb, but once persuaded by a certain soldier to eat, she is soon enough copulating on the spot. There is a certain suggestion of the widow of Ephesus in Hamlet's mother, who in her own way went from grief to sex with shocking facility, and served the leftovers from King Hamlet's funeral at the wedding that followed close on its heels (or so Hamlet sarcastically alleges). For the court it is as if grief over the king's death were washed away by the communal rituals of feasting and coronation. It is otherwise with Hamlet. Hamlet's melancholy defines itself as *not* being subject to the sorts of social remedies recommended by tradition and catalogued by Burton, such as conviviality. Passing quickly is exactly what his gloomy lamentation does not do.

↹

"The quality of social ties is essential to health outcomes."[3] In so rating our bonds to others, medical and social science follow knowingly or unknowingly in the footsteps of a long tradition holding, similarly, that company itself is a kind of medicine—drawing us away from our sorrows—and that too much solitude can harm. The physician Rabelais tells us that he writes his comic chronicles for no other end than to help the sick and the sorrowful, as if literature itself could provide cheering company. Of his "pantagruelic mythologies" he reports that "many of the ailing, the sick, the weary or the afflicted have, when they were read to them, beguiled their benighted sufferings, passed their time merrily and found fresh joy and consolation."[4] If the works of Rabelais constitute a pharmacopoeia of joy, Burton's *Anatomy of Melancholy* is an encyclopedia of sorrow—including, however, the remedies of sorrow. Burton's converse of the proposition that the less socially integrated are more likely to die is "the merrier the heart, the longer the life."[5]

In a digression on spirits as a cause of melancholy, Burton mentions "devils or the souls of damned men that seek revenge, or else souls out of purgatory that seek ease" (I.193-94): a company that recalls the Ghost in *Hamlet* both insofar as Hamlet suspects it of seeking *his* damnation and insofar as the Ghost seems to dwell by day in purgatorial fire—and it is characteristic of the play that it calls up not one or the other but both of Burton's alternative possibilities. But even before he meets the Ghost, Hamlet suffers from melancholy to the point of yearning for an end to his existence. That the Ghost simply compounds Hamlet's already unbearable woes suggests that Hamlet's melancholy is beyond the reach of the social antidotes passed down by tradition.

According to Proverbs, "A merry heart doeth good like a medicine" (17:22). Recorded by Burton are many remedies of this kind—social measures that produce a medical effect without the use of a medicine, in the manner of a placebo. (As we might say, in these cases social factors do not provide an activating context for a treatment, but are themselves the treatment.) All liberal physicians, reports Burton, "will have a melancholy, sad, and discontented person make frequent use of honest sports, companies, and recreations" (II.121).[6] Merriment relieves sorrow; it is balm for the soul. "Merry company is the only medicine against melancholy" (II.124); "company, a sole comfort, and an only remedy to all kind of discontent" (II.125). In Shakespeare's tragedy of melancholy, however, such approved social antidotes seem only to aggravate the hero's disgust with life and self—carousing sessions, for example, drawing his contempt, and the merry company of Rosencrantz and Guildenstern connoting servility and betrayal. I propose consulting Burton not so much for the light he may shed on Hamlet's malady (an intentionally bewildering compound of grief, disgust, pretense, and mirth in the loss of mirth itself) as for the light he does shed on the king's attempts to cajole him out of it by means of ordinary social expedients. Conventional remedies for melancholy are recommended for one whose melancholy does not arise from ordinary sources such as bereavement on the one hand or rejection in love on the other—again characteristically, Hamlet endures both—and who plays with and on conventional poses associated with the disease, from the aggrieved lover to the satirist and the malcontent.[7] The king's stock approach to Hamlet's malady casts doubt on the quality of his concern and helps explain why his various prescriptions do not work or even work in reverse.[8]

The last words of Burton's stupendous tome, thematic of the whole, are "Be not solitary, be not idle,"[9] which is no more than common sense. Both common sense and our affiliation with others also tell us that we belong to the human race and our misfortunes are therefore the common lot of human life, however sharply we feel them. Asks one of Burton's authorities, "If it be common to all, why should one man be more disquieted than another?" Or as Burton himself says, "If thou alone wert distressed, it were indeed more irksome, and less to be endured; but when the calamity is common, comfort thyself with this, thou hast more fellows. . . . 'Tis not thy sole case, and why shouldst thou be so impatient?" (II.128) Hence Gertrude's reproof of Hamlet, "Why seems it so particular with thee?" (1.2.75)[10]—a point repeated in short order, less gently, by the king. At the heart of *Hamlet* Frank Kermode finds an "evil doubling,"[11] and these rebukes of Hamlet delivered in full view of the court reproduce twice over the argument that "if it be common to all" one should not make too much of sorrows, but with the repetition of the lesson serving to defeat its intent by reminding Hamlet so forcibly and painfully that his mother and uncle are in fact one. In the spirit of homily, both Gertrude and Claudius reduce his grief to anti-social behavior. "If it be common to all" becomes for Gertrude, "Thou knowest 'tis common. All that lives must die," and for Claudius,

> For what we know must be, and is as common
> As any the most vulgar thing to sense,
> Why should we in our peevish opposition
> Take it to heart? (1.2.98-101)

Likewise, "Why shouldst thou be so impatient?" becomes Claudius' "A heart unfortified, a mind impatient, / An understanding simple and unschooled" (1.2.96-97). Burton is a great anthologist of commonplaces. Presented with a young man dressed in the garb of melancholy, both Gertrude and Claudius invoke a commonplace antidote to that malady.

The intent of social remedies for melancholy is that the good spirits of others should rub off on the sad one, that he or she should feel what others feel instead of suffering alone. It is in this sense that merriment relieves sorrow. So it is that while he cannot make Hamlet join in the festivities in the castle in Act One, the king does conscript him symbolically by declaring that in honor of Hamlet, "No jocund health that Denmark drinks today / But the great cannon to the clouds shall tell" (1.2.125-26). In effect, Hamlet is to be cheered up in absentia. While "health" here evidently means "toast," there may be a suggestion that festivity, jocundity, itself promotes wellbeing, much as Burton and his authorities say. As noted, also approved by Burton as a remedy for melancholy is "a cup of good drink now and then" (II.120). But if the selfsame reproach delivered by his mother and his new father serves only to embitter Hamlet's melancholy (his soliloquy a few lines later begins with thoughts of suicide and dwells on his mother's infidelity), so too the drinking bouts in the castle excite his repugnance and contempt. They are "a custom / More honoured in the breach than the observance" (1.4.17-18; cf. 3.2.277).

It is after Hamlet agrees to remain in Denmark that Claudius rises to drink and celebrate. Though Hamlet accedes to his mother's request and pointedly not the king's, his compliance nevertheless gratifies the king. But why did Gertrude and Claudius ask or tell him not to return to Wittenberg in the first place? As for Gertrude, perhaps it is enough to say that a mother loves her son and desires his company. The king's motivation seems more shady.

> For your intent
> In going back to school in Wittenberg,
> It is most retrograde to our desire,
> And we beseech you, bend you to remain
> Here in the cheer and comfort of our eye,
> Our chiefest courtier, cousin, and our son. (1.2.112-17)

Does the king want to keep an eye on Hamlet? Burton does advise keeping the secretive melancholic under observation. "If he conceal his grievances, and will not be known of them, '[friends] must observe by his looks, gestures, motions, phantasy, what it is that offends'" (II.110), counsel that gives a semblance of justification to the king's attempts to spy on Hamlet and figure him out. Just so, "cheer and comfort" will be good for Hamlet. What Hamlet in his state of despondency really needs is to be in the company of happy others, in particular the king. But one doubts the king's show of therapeutic concern. Melancholy, according to Burton, signifies "a discontented and troubled mind" (II.126), which in a displaced heir apparent bears watching. Keep your friends close and your enemies closer. A related reason for removing Hamlet from his studies is also to be found in Burton: deep study itself conduces to melancholy. If Hamlet were to return to Wittenberg, therefore, he might abandon himself not only to his grief but his discontents and grievances. "Avoid overmuch study and perturbations of the mind . . . Amidst thy serious studies, use jests and conceits, plays and toys, and whatsoever else may recreate thy mind" (II.124).[12] Later in the play Claudius orders Hamlet out of Denmark; here, with what seems to be misplaced confidence in social remedies for melancholy, he detains Hamlet in Denmark when Hamlet would have willingly departed for Wittenberg.

The king's next move, seconded by the queen, is to summon Hamlet's friends Rosencrantz and Guildenstern, seemingly with the same double motive of keeping him under observation and cheering him up.

> I entreat you both
> . . . by your companies
> To draw him on to pleasures, and to gather
> So much as from occasions you may glean . . . (2.2.10, 14-16)

If friendship is medicinal—an ointment according to Plutarch, "the medicine of life" according to Cicero—by the same token false friendship is poisonous,[13] as Hamlet's imaging of Rosencrantz and Guildenstern as fanged adders later in the play strongly suggests.

Given that keeping an eye on Hamlet while surrounding him with "cheer and comfort" hasn't worked so far, it is hard to see why the same tactic should work via Rosencrantz and Guildenstern. That the king believes or wants to believe in social remedies for melancholy is suggested by his strong response to Rosencrantz's report that "a kind of joy" (3.1.19) seemed to spring up in Hamlet at the news that the players were en route to Elsinore:

> . . . it doth much content me
> To hear him so inclined.—Good gentlemen,
> Give him a further edge, and drive his purpose on
> To these delights. (3.1.25-28)

A melancholy that brightens at the prospect of a play would seem to be an ordinary indisposition, responsive to ordinary therapies: a thought that brightens the king. If Hamlet can be entertained out of his melancholy, then a disturbance of the king's peace of mind and threat to his secure enjoyment of power will have been quelled. Perhaps Hamlet is simply catching up with the rest of the kingdom of Denmark, which some weeks ago went from grief over the death of King Hamlet to the joy of his brother's marriage and coronation. After all, as we learn in the central speech of the Mousetrap, grief turns to joy on "slender accident" (3.2.18). The use of theater to cheer a melancholy mind is entirely consonant with Burton's prescription of "scenical shows" (II.123) and liberal recreations.

As if he were subject to the agency of others and not an agent in his own right, Hamlet is to associate with Rosencrantz and Guildenstern but not with Ophelia, and is sent to England—as a remedy for melancholy—after being told not to return to Wittenberg.

> There's something in his soul
> O'er which his melancholy sits on brood,
> And I do doubt the hatch and the disclose
> Will be some danger; which to prevent,
> I have in quick determination
> Thus set it down: he shall with speed to England
> For the demand of our neglected tribute.
> Haply the seas and countries different,
> With variable objects, shall expel
> This something-settled matter in his heart
> Whereon his brains still beating puts him thus
> From fashion of himself. (3.1.163-74)

According to Burton, "no better physic for a melancholy man than change of air and variety of places, to travel abroad and see fashions" (II.67). At this point, however, Claudius no longer puts on much of a show of concern for Hamlet; the brooding young man is to be sent packing to rid Denmark of a certain danger. Whether or not Hamlet's melancholy is relieved by a change of air and scene, shipping him to England will get him out of the king's way.

Once the king knows, via the Mousetrap, that Hamlet somehow knows or suspects the truth of his father's death, he (the king) must act, and it may be at this point that he decides that Hamlet is not only to be sent to England but summarily executed on arrival. The simple fact that he sends Hamlet away, whereas in Act One he prevented Hamlet from leaving Denmark of his own accord, shows that his efforts to coax Hamlet out of his melancholy by various social measures have come to nothing and, in fact, were misconceived. Just as his decision to keep Hamlet at hand proved self-defeating, and just as his hopes for an evening of theatrical entertainment backfired, so his policy of beguiling Hamlet out of his solitude by the use of social distractions simply drove Hamlet into himself and gave a further edge to his grief and rage. Hamlet's many sudden turns from merriment to cruelty and contempt might be taken as a way of counterattacking the therapy of happiness that has been prescribed for him.

In this Hamlet resembles (with allowance for differences) the dialogical heroes of whom the literary theorist M. M. Bakhtin has said that they furiously contest definitions of themselves by other people, that they try to "outguess and outwit" these others, and in their vital indeterminacy defy the formula of identity, A = A.[14] At once mad and not mad, playing hostile games of wit with those seeking to understand and manage him, Hamlet too seems to be resisting definitions, and the definition most readily available for the young man who presents himself in black is the melancholic. This is not to say that Hamlet isn't grief-stricken, but that his malady has more and deeper sources than others imagine, just as it seems to spring all at once from the disastrous events that have befallen him and from something deep within.

As determined by the American Psychiatric Association, the normal period of mourning following the death of a loved is two months, with symptoms of depression that persist beyond this cut-off point ineligible for the "bereavement exclusion." By the middle of *Hamlet* we learn that King Hamlet has been dead "twice two months," so that by statute Hamlet's grief should have ended—which only establishes the irrelevance of arbitrary standards to his case. While conventional remedies for melancholy are underwritten by tradition rather than professional authority, they fit him and his circumstances no better than the bereavement statute. The remedies may well be appropriate for most, but because Hamlet's case is so far from ordinary, in fact so unlike any other on record, such generic measures are bound to fail. Burton himself, the cataloguer of approved remedies, underscores the singularity of cases, just as he recognizes that melancholy overflows any system employed to classify and define it.

> The four-and-twenty letters make no more variety of words in divers languages than melancholy conceits produce diversity of symptoms in several persons. They are irregular, obscure, various, so infinite, Proteus himself is not so diverse. (I.408)

Now dilatory, now rash, the lover and the reviler of Ophelia, Hamlet himself is irregular, obscure, ever-changing, and seems to have enough causes and sources of melancholy for more than one person. Even before he learns of the assassination of his father, the glaring indecency of his mother's hasty marriage to King Hamlet's brother—an indecency that somehow makes no impression on the court—is enough to flood him with disgust and despair. To this is added the shock of meeting the ghost of his murdered father, now suffering the pains of Purgatory, and being commanded to carry out a mission of revenge that demands a single-mindedness seemingly contrary to the diversity of his nature.

Over the course of the play, at least until his departure for England, Hamlet's melancholy becomes a vortex that seems to suck in everything including the speeches that are his very signature, the soliloquies. The soliloquies constitute explicit violations of Burton's advice not to be solitary, not to be idle. In order to utter a soliloquy you have to be solitary, and Hamlet's speeches of this kind in the middle acts of the play not only reflect on his idleness but complicate it by serving, perversely, as substitutes for action. It is as if melancholy had become self-propelling, as if his speeches not only expressed disgust with life and self but doubled it and made it more unbearable. It is no common ailment that feeds on itself this way.

"When the calamity is common, comfort thyself with this, thou hast more fellows." But Hamlet's calamity is actually one horror after another, a wave of them. He has no fellows. His tragedy begins in his singularity, which in and of itself rules out social consolations. So it is that efforts to cheer and distract Hamlet and draw him into the circle of social life have exactly the opposite effect: they sting him, turn his mind on itself, and alienate him still further from others. To the already too-numerous sources of his melancholy another is added.

Burton indeed recognizes a certain ironic potential in his prescriptions. If merriment relieves us and distracts us from our suffering, should we then make it our occupation?

> But see the mischief; many men, knowing that merry company is the only
> medicine against melancholy, will therefore neglect their business, and, in
> another extreme, spend all their days among good fellows in a tavern or an
> ale-house, and know not how to bestow their time but in drinking . . . like
> so many frogs in a puddle. (II.124)

In a scene that lives in the world's memory as the emblem of *Hamlet*, two
commoners with the most melancholy of occupations turn a graveyard into
a tavern, each necessary to the other not for labor but profane fellowship,
frogs in a puddle. Taking in this spectacle with Horatio, Hamlet is offended
by their clowning, in part perhaps because it recalls so crudely the "mirth in
funeral" (1.2.12) with which the current king began his reign.

From the denial of the last rites to King Hamlet to the celebration of a
marriage close upon a funeral to the unceremonious burial of Polonius and
the brawl in Ophelia's grave, *Hamlet* tells of violated sacraments and rituals.
However, not only formal observances but even the informal ceremonies of
social life seem broken in *Hamlet*. Song is maddened, conversation vexed,
letters returned, merriment embittered, drink poisoned. The bonds that constitute human society and make it a source of sustenance and consolation, as
in Burton, have failed. It is as if it had been reduced to a collective fiction. A
recent meditation on beauty observes in passing that "conventions create a
background of unchanging order in our lives, a sense that there is a right way
and a wrong way to proceed."[15] With the destruction of the order of things
as Hamlet recently knew it, and the concurrent collapse of the distinction
between the right and wrong ways to do things, convention itself now seems
corrupted. The conventionalism of the king's remedies and recommendations
for Hamlet's melancholy suggests that his concern for Hamlet is merely a
social fiction, a mask.

Coincidentally or not, it is when the king is unmasked in front of all, first
by Gertrude ("The drink, the drink—I am poisoned"), then Laertes ("Thy
mother's poisoned. . . . The King, the King's to blame"), that Hamlet finally
kills him. But the revelation that the wine as well as Laertes' sword have
been poisoned also exposes the pretense that the fencing match itself is one
of the "honest and chaste sports . . . , plays, games" recognized as a remedy
of melancholy (II.123). Indeed, Burton specifically includes fencing among
the physical exercises and pastimes said to cure melancholy (II.74). That the
king should be caught in a death-trap of his own devising is poetically fitting,
but it is no less fitting that Hamlet's revenge should be carried out at last in a

tainted proceeding masked as one more innocent social diversion.

The placebo effect is in good part socially driven, and the various approved remedies for Hamlet's melancholy proposed, sincerely or not, over the course of the play—companionship, merriment, distraction, travel, games—are all social in nature. They all backfire. A play imbued with skepticism may also constitute the first thoroughly skeptical analysis of the placebo effect on record.

Chapter Four

The Power of Suggestion
Eve and the Apple

For a moment Eve experiences the exaltation she was led to expect.

The placebo effect conventionally involves the attribution of therapeutic properties to something that does not have them. According to *Paradise Lost*, the very history of humanity turns upon the attribution of psychotropic powers to an ordinary object—an apple.

⊟

In an attempt to injure God by attacking his creatures, Satan makes his way to the Garden of Eden, where, using the form of a serpent as his Trojan Horse, he has the good fortune to find Eve alone. Conducting her to "the tree of prohibition," he rises to the height of false eloquence, extolling the tree and its wonder-working power, flattering Eve, and arguing that she has been unjustly confined to her lowly station by a tyrannical "Threat'ner." "O sacred, wise, and wisdom-giving plant," he exclaims,

> Mother of science [knowledge], now I feel thy power
> Within me clear, not only to discern
> Things in their causes, but to trace the ways
> Of highest agents, deemed however wise.
> Queen of this universe, do not believe
> Those rigid threats of death; ye shall not die:
> How should ye? By the fruit? It gives you life
> To [that is, in addition to] knowledge. By the Threat'ner? Look on me,
> Me who have touched and tasted, yet both live,
> And life more perfect have attained than fate
> Meant me, by vent'ring higher than my lot.

The very audacity of the serpent's claim may recommend it, in accordance with Hume's observation that quacks and other illusion-peddlers who make bold claims "meet with a more easy faith upon account of their magnificent pretensions, than if they kept themselves within the bounds of moderation."[1] In this case the audacious claim is all the more persuasive in that it is presented as something tried and proven. Satan does not encourage Eve to do what no one has attempted before, which would have scared her off, but simply to repeat what he has already done with success and impunity. "Look on me." Moreover, the very act of making an argument proves, or seems to prove, the power of the apple, in that his ability to speak supposedly came from the apple itself—a Miltonic innovation.[2] In citing his own case as a precedent he raises the expectation of wondrous effects to be enjoyed (effects demonstrated and certified by his rhetoric) and gives Eve a model on which to pattern her own experience of the apple if she can be brought to eat it.

Intuitively Satan has hit on a mechanism of the placebo effect, which we do not experience at random but specifically where we are led to expect it. Such is the influence of expectation on the experience of our physical selves that study subjects not only report more pain when led to expect it, even when no pain at all is administered, but report less of the pain-killing effect of morphine when they do not see the drug being administered and thus do not know its effect is on the way.[3] Often, however, it is because of the example or seeming example of others that we form an expectation in the first place. We feel what we suppose others feel. Satan offers his own experience of the apple's transforming effects—"[N]ow I feel thy power"—as just such a model for Eve.

Seduced by serpent's fictitious example and emboldened by his eloquence to defy the one prohibition binding on her and Adam, Eve succumbs in short order and eats the apple. "Earth felt the wound."

No sooner does Eve taste the fruit than she feels or imagines she feels its power, just as she was led to expect. The pursuit of higher knowledge leads to an immediate surrender of spontaneity:

> Eve
> Intent now wholly on her taste, naught else
> Regarded, such delight till then, as seemed,
> In fruit she never tasted, whether true
> Or fancied so, through expectation high
> Of knowledge, nor was Godhead from her thought.

Like the serpent, she proceeds to apostrophize the tree itself:

> O sovran, virtuous, precious of all trees
> In Paradise, of operation blest
> To sapience, hitherto obscured, infamed,
> And thy fair fruit let hang, as to no end
> Created. . . .

The word "virtuous" here is the adjectival form of "virtue" in the sense of "power to affect the human body in a beneficial way" (Oxford English Dictionary), just as we might say of a drug. Satan infiltrated the garden of Eden on a mission of fraud, and fraud is now coursing through Eve, convincing her that her mind has been uplifted through the operation of a fruit that acts like a metaphysical medicine. Even if she hadn't been shown making an idol of a tree, we would have known from Milton's language that her mind has fallen captive to its own expectations and inflated imaginings. "Whether true or fancied so." Following immediately upon the serpent's eulogy of the apple, the intoxicating action of the fruit dramatizes the placebo effect.

If she is to experience a feeling of transcendence, Eve must believe that she is responding to the properties of the apple, not just to the serpent's suggestions and promptings.[4] So it is that in her address to the fruit she does not even mention the serpent—a telling absence noted by an early commentator:

> Our author very naturally represents her in the first transports of delight expressing her gratitude to the fruit, which she fancied had wrought such a happy change in her, and next to *experience* her *best guide* [lines 807-08]: but how is it possible that she should in these rapturous acknowledgments forget her guide and instructor the Serpent, to whom in her then notion of things she must think herself the most indebted? I don't doubt that Milton was sensible of this, but had he made Eve mention the serpent, he could not have avoided too making her observe that he was slunk away, which might have given her some suspicions, and would consequently have much alter'd the scene which follows betwixt Adam and her.[5]

Had she mentioned the serpent she would have noted his absence and might have drawn conclusions. But there is another, perhaps stronger reason the serpent does not figure in Eve's reverie. She can't afford to think of the serpent; to do so might mean realizing that she is not responding to the apple but to him, that her sensation of transcendence is not only imitative but fictitious, and that she risks bringing death on herself and her descendants for precisely nothing. Patterning her response to the apple on the model provided but denying that she is doing so, Eve compounds the serpent's deception of her with deception of herself.

The "first transports of delight" do not last. Being fallacious, Eve's experience of exaltation dissipates almost immediately, to be succeeded by one desperate imagining after another: that she has perhaps escaped divine observation (as the retired serpent has escaped hers), that the apple will remedy her galling inferiority, that Adam will wed "another Eve" if she should suffer death. Consider her notion that with luck her crime went unnoticed, a drastic fantasy poles apart from transcendent knowledge:

> Heav'n is high,
> High and remote to see from thence distinct
> Each thing on earth; and other care perhaps
> May have diverted from continual watch
> Our great Forbidder.

Like the idea that God can somehow be hoodwinked, the epithet used for God is a product of deceit and self-deceit; it mimics the serpent's defamation of God as "the Threat'ner" in his address to Eve. Here and elsewhere, the text shows us that Eve's experience of the apple's elevating effects is both fictitious and plagiaristic,[6] and that these vicious qualities have much to do with one another. But what if God has not been somehow distracted? In that case death will ensue.

> Then I shall be no more,
> And Adam wedded to another Eve,
> Shall live with her enjoying, I extinct;
> A death to think. Confirmed then I resolve,
> Adam shall share with me in bliss or woe.

If Eve is to die, so must Adam die; misery loves company.

Seeking to "discern / Things in their causes" and anatomizing the workings of the mind, Milton takes a skeptical view of the placebo effect resembling the understanding arrived at by science a century and more later. Yet he was probably not interested in advancing understanding of the production and consumption of medical illusions—though we can well imagine such a concern on the part of his contemporary Locke, a physician keenly aware of the mind's fallibility. From where, then, arose Milton's sense of the placebo effect as a fallacious experience begotten by expectation?

My sense is that it flowed from his religious convictions, specifically his abhorrence of idolatry.[7] In attributing magical power to an apple and actually adoring the tree it grows on, Eve commits this capital sin. That the fruit is imagined not just as mysteriously potent like any object of superstition, but as containing the power of divinity itself, makes its worship all the worse. Both superstition and idolatry were strongly associated by Milton with Catholicism, the arch-enemy of human dignity and liberty. His "Areopagitica," an impassioned plea for more or less unlicensed printing that vilifies censorship as a Catholic invention, may therefore cast some light on the operation of the placebo effect in the garden of *Paradise Lost*.

"Truth is compared in Scripture to a streaming fountain; if her waters flow not in a perpetual progression, they sicken into a muddy pool of conformity and tradition." For Milton Catholicism is not merely a blind attempt to imprison the infinite and chain the wind, but a system of conformity and enslavement. "How goodly and how to be wished were such an obedient unanimity," he exclaims sarcastically, "what a fine conformity it would starch us all into." Even as he rejoices in the second coming of the Reformation, he fears a national relapse into "a gross conforming stupidity."[8] The placebo effect as enjoyed by Eve in *Paradise Lost* is portrayed, accordingly, as an essentially derivative experience, an artifact of conformity. Eve patterns her exaltation on the exact template the serpent was careful to provide, thereby providing an instance of the willing subjection decried by Milton as the bane of spiritual liberty. Because the serpent has tasted the apple and attained a transcendent knowledge previously denied him, and has gotten away with it (so he claims), Eve momentarily imagines herself doing the same.

Perhaps it is because idolatry spreads, replicates itself, in this way that in the first book of *Paradise Lost* idolatry is portrayed as a corruption that engulfed "the greatest part / Of mankind." One mode of idolatry somehow suggests or metastasizes into another, so that to enumerate them all requires nothing less than a catalogue (which the poet supplies), though even that will be incomplete. As an extant system of idolatry, Catholicism as Milton understands it not only partakes of servile imitation but breeds it, as the crypto-Catholic attempt to regulate printing in England imitates the Spanish Inquisition. Eve's imaginary enjoyment of the apple's power is itself, just so, an imitative act. The only "wisdom-giving" power contained in the apple is the aftertaste of remorse.

The spreading potential of corrupt practices—the tendency of one imitation to beget another—is dramatized in this episode itself as Adam hears at second hand the serpent's claims about the apple's "divine effect" (now allegedly confirmed by Eve's experience), repeats them, talks himself into anticipating an "ascent" to a higher form of being, and follows Eve's precedent and eats the apple, whereupon Eve herself does so a second time. Justified by the same lies, they abandon themselves to the same fictitious experience of exaltation:

> As with new wine intoxicated both
> They swim in mirth, and fancy that they feel
> Divinity within them, breeding wings
> Wherewith to scorn the earth.

They fancy they feel what the serpent reported he felt. As Adam eats, the poet tells us that he does so "Against his better knowledge, not deceived." Remarkably, then, Adam imagines, at least for the moment, that he feels an intoxicating divinity within him even though he knows full well that the apple contains no power. At just this point the divine "virtue" of the apple vanishes like the fiction it is, to be replaced by sheer lust, as if the two were inspired not by any property of the fruit itself but simply by their own dishonesty in seizing it. At this point each indeed feels what the other does. "He on Eve / Began to cast lascivious eyes, she him / As wantonly repaid."

The apple contains no magical property, no "sciential sap." The knowledge of evil it confers is nothing but the remorse that follows the crime of eating it. It is because the apple is exactly what it seems to be, simply an apple, that its elevating effects quickly wear off both when Eve eats the fruit alone and when she and Adam do so jointly in a lewd feast.

One of the first scientific writings in England on the placebo effect, at the end of the eighteenth century, made a point of the transience of placebo benefits. "I have sometimes observed," states the author, John Haygarth, "that the administration of a new medicine . . . if recommended to the publick with exalted praise, has been attended with great success—much greater than what was confirmed by future experience."[9] In his experiments on the placebo effect, Haygarth made sure to raise the expectations of the study subjects, as the serpent did with Eve, before treating them with inert objects. Haygarth did not follow Milton, either consciously or unconsciously. However, by exposing the falsity of certain "wonder-working" devices,[10] he did advance the disenchantment of the world to which the Puritan movement contributed both negatively by the destruction of stained glass and positively by its affirmation of "ordinary life."[11]

The Puritan ennoblement of ordinary life entailed such principles as the dignity of labor, the denial "that there are special places or times or actions where the power of God is more intensely present,"[12] and the dignity of marriage—the last point illustrated in the concord of the unfallen Adam and Eve. The portrayal of the apple as no magical thing, but simply an apple, is entirely consonant with the Puritan affirmation of the ordinary. As it happens, some of the first experiments on the placebo effect in France took place in the gardens of an American of Puritan descent whose writings and thought are imbued the importance of the ordinary: Benjamin Franklin. In a variant of the principle that divine power does not manifest itself in special ways, Franklin and his colleagues demonstrated that trees theoretically magnetized by a disciple of Mesmer could not be distinguished from an ordinary tree, and those who thought they could do so deceived themselves.

According to the still unfallen Adam, "To know / That which before us lies in daily life / Is the prime wisdom." Seeking wisdom in an apple, Eve mistakes the nature and the locus of wisdom; aspiring to become a god, she seeks to remove herself from the realm of the ordinary altogether. The quotidian character of Franklin's wisdom is well known to the world. "Early to bed and early to rise, makes a man healthy, wealthy and wise." (Poor Richard also says, "God heals, and the doctor takes the fees."[13]) It is through his affirmation of "daily life," then, that a poet not especially concerned with the evaluation of medical claims and the verification of scientific knowledge was able to portray the operation of the placebo effect with unprecedented subtlety and anticipate its investigation a century later.

Chapter Five

To Feel What Others Feel
Two Episodes from 18th-Century Medicine

The first investigations of the placebo effect were in response to dubious therapies that caught the public imagination.

In an essay of 1750, Samuel Johnson noted that improbable tales of shipwrecks and captivity, of castles and ogres, had gone out of favor with a reading public that now expected stories to observe the boundaries of everyday life. This shift from the fantastic toward something like common sense represents more than a passing fashion, he implies, and is a good thing (analogous perhaps to the overcoming of barbarous usages in other domains) even if it makes work more difficult for the novelist—more difficult in particular because he is now responsible for showing inexperienced readers how to navigate the trials and tests of daily life itself. According to Johnson, readers of fiction read precisely in order to learn how to live. "When an adventurer is leveled with the rest of the world, and acts in such scenes of the universal drama as may be the lot of any other man, young spectators fix their eyes upon him with closer attention, and hope, by observing his behavior and success, to regulate their own practices when they shall be engaged in the like part."[1]

Nevertheless, Johnson knew that his contemporaries also read fiction for pleasure and because they were interested in the fortunes of others, especially others like themselves. But how is it that after being for so long focused on figures of a higher order than those presented to us in everyday life—figures like the heroes of epic, tragedy and romance—fiction now found readers so responsive to characters "leveled with the rest of the world"? Some might say that the time for Tom Jones had arrived, or that the flow of history or literary history dictated that archaic styles give way, or that the bourgeois reading public demanded a fiction constructed in its own image. However that may be, the fiction of everyday life also both drew upon and fed the belief that the experiences of others are much like our own, quite as if they and we had the same sensations. Belief in the uniformity of human sensations is not one that comes readily to all people in all times and places. In the decades following

Johnson's paper it was connected to a new persuasion that we experience as others do because we are constituted as others are. The presumed sameness of our nature and physical nature enables us to feel what others feel, both morally and even physically, and attracts us to their experiences. The possibility of such an identity with others was itself attractive, and perhaps the more so because of its novelty.

The proving ground for ideas regarding the human body and its susceptibilities is medicine. Traditional medicine, while it certainly employed standard practices like bloodletting and the use of purgatives, tailored diagnosis to the specific constitution of the patient—in theory at least. The Hippocratic writings, it is said, display a strong awareness of "the individuality of each patient," in part because "for many authors, the difficult part of medicine is being able to judge precisely what is happening in each patient, to distinguish the individual from the general."[2] Galen would have agreed, for although he "recognised groups of symptoms as constituting diseases, these offered only general indications of what was wrong, and every patient suffering from, say, apoplexy or dropsy had an individual balance that needed to be restored. ... The true doctor tailored his medicaments to the individual patient." Medieval medicine, in the general tradition of Galen, was complexional, "but since complexion differed in each individual, a really satisfactory health regime would have to be tailored to individual needs."[3] (Perhaps this is why medications in the medieval Islamic world seem to have employed many ingredients in complicated combinations.) Chaucer's Physician performs certain astrological calculations "for his pacient," a phrase implying that he too adapts treatment to the particulars of the case, or at least wants to be seen to do so. Only then does he send to the apothecary. It is unlikely that all of Galen's patients with dropsy or all of the Physician's patients with, say, catarrh would have recognized themselves as experiencing the same thing, even assuming they were in communication. In the decade of Johnson's death a notable or notorious medical fashion was animated precisely by the principle that all are subject to the same experiences, that our ailments trace to one and the same cause, and that the same universal force acts on us. Fanned by pamphlets and the press, the fashion spread as more and more people felt what they supposed or imagined other devotees felt.

The fashion was Mesmerism. The fascination of France in the twilight of the pre-revolutionary era, Mesmerism was the brainchild of a medical doctor convinced that a fluid called animal magnetism suffused all things and that ailments of the human body were caused by blockages of this power which could be cleared by his ministrations. Fixing subjects with his eyes and massaging various magnetic poles on their person, arrayed in robes, accompanied by the ethereal tones of a glass harmonica that certainly contributed their share to the power of context, the maestro induced "crises" by discharging the elemental power that had somehow been dammed up. Much as vats of iron filings and magnetized water served to dramatize the allegedly physical nature of the elemental fluid, so convulsions and other unusual performances gave physical form to the cathartic experience of Mesmer's subjects. But if the expressive drama of Mesmerism was sure to excite the public imagination, arguably the arrow of causality goes the other way too: only a therapy that had hold of the public imagination could have produced such vivid results in the first place.

If, as some believe, psychotherapy traces to Mesmer, the treatments for which he is best remembered were not conducted in strict privacy but, on the contrary, with a theatricality designed to impress all who witnessed or perhaps even heard of them.[4] After all, animal magnetism operated in and on everyone, whether participant, spectator or reader. The role of print in stirring public fascination with Mesmerism bears emphasis, for no sooner were cures performed than they were written up in pamphlet form to join a multitude of other writings as well as newspaper accounts without which the Mesmer movement would not have been a movement. "Everyone is occupied with Mesmerism," remarked one observer—in print.[5] While the ability of Mesmerism to be everywhere and stir everyone must have seemed like an illustration of the fundamental power itself, it was print that gave Mesmerism this omnipresence. Print itself, like the theorized magnetism, was a medium charging and suffusing all things. Print enables people who are not present to one another—as partakers of the same drug (in Homer) or citizens engaged in identical rituals (in More) *are* present to one another—to participate in common experiences nonetheless.

Inspired as it was by the ideal of overcoming obstruction and restoring the harmony of humanity and Nature, Mesmerism had a distinctly Rousseauvian resonance. Indeed, Mesmer discovered Mesmerism after wandering in a forest for three months "like a Rousseauite savage,"[6] communing with Nature the better to clear his mind of bad ideas. But Rousseau was a novelist as well as a polemical theorist, and trances and raptures of a sort are also available to readers of fiction, perhaps especially readers "easily susceptible of impressions" and "open to . . . false suggestion," as Johnson put it and as critics of Mesmerism would in turn say of those who came under its spell. Even now "mesmerizing" is a term of praise in blurbs and reviews. The scientific philosophy underlying Mesmerism—that of "sentimental empiricism," the doctrine "that feelings were responses to a world outside the mind and were therefore the bedrock of natural knowledge"[7]—likewise seems congenial to a literary form with a strong potential for the sentimental and a strong empirical interest in things "that daily happen in the world," in Johnson's words. Perhaps those who read about Mesmerism, and they were many, had an experience similar to readers of a tale of credible yet remarkable events.

A few years after Mesmerism went into eclipse with the onset of the Revolution, another medical craze broke out—this time in the United States, and secondarily in England—in which bodily ailments were treated by the channeling of a theorized elemental fluid, in this case animal electricity or something like it. As a result of empirical trials, apparent successes, and publicity proclaiming the new treatment as simultaneously remarkable and credible, the idea caught on that certain patented rods known as tractors (for their power to draw things from the body) were capable of curing everything from ordinary aches and pains to epilepsy if simply passed over the subject. Invented by the physician Elisha Perkins, the tractors were championed after his death by his son Benjamin, who indignantly denied that their operation bore any similarity to Mesmerism, a practice he regarded as charlatanry pure and simple. Clearly, however, these were kindred phenomena. It is not just that the Perkins rod, like Mesmer's, was wielded like a wand, and that both evoke the lightning rod, and both were said to be in touch with a physical force too profound and subtle for ordinary perception. More fundamentally, and in contrast to a medical treatment specific to a given patient, both therapies address themselves to humanity in general (employing a rhetoric of philanthropy), and to ailments of all kinds. Both presume that a single force is at work in all bodies, somehow responsible for wellbeing or illness, and both activate this presumption of universality by inviting those in search of healing to experience the same thing that many others have already experienced.[8]

In the case of the Perkins tractor the invitation to feel what others feel was all but explicit. If Mesmerism attracted followers as though magnetic itself, Benjamin Perkins spread word of his father's invention by citing the testimonials of respected persons one after another, all documenting the sort of benefits to be reaped by anyone who may use the tractor. Dr. Rufus Johnson of Connecticut reports, "In the course of my practice a few months past, I have made frequent experiments with the Metallic Tractors, and have, with but very few exceptions, succeeded to my surprise in removing rheumatic pains, head-achs, pains in the face, spasmodic affections, and inflammatory swellings of the throat." Dr. Thomas Backus of Connecticut attests that he treated an epileptic patient with the tractors and "in less than one minute he was entirely free from every kind of spasm, and instantly regained his reason. I still continued by his side, and prevented several other general attacks, by operating at the beginning of the symptoms." Dr. Samuel H. P. Lee, also of Connecticut, attests that "a lady fell from her horse and dislocated her ancle, which remained several hours before it was reduced, by which it became very much swelled, inflamed, and painful. Two or three applications of the Tractors relieved the pain, and in a day or two she walked the house, and had no further complaint."[9] Presumably it was the circulation of reports like these, seconded by word-of-mouth recommendations, that drove the Perkins movement. The implication of the many reports cited by the younger Perkins is not that the tractor requires expert hands (after all, it is simply waved over the patient) but that so many affidavits by so many credible authorities prove the instrument's power—and that no limit can be set to that power's application. The sort of benefits many have enjoyed, many others can also enjoy. With marshaled testimony serving as a kind of publicity machine and prospective users of the tractor invited to model their experience on that of others whose stories are recorded, it is as if the benefits of "Tractoration" came about not so much through the agency of electricity as through the also universal medium of the word—in any case, through social rather than medical means.

Though admitting that he could not say how his tractor worked, Perkins maintained that repeated experiments had proved beyond a doubt that work it did.[10] "It is a favourite maxim with Dr. P.," writes his son, "that the judicious physician at first leaves the flowery path of speculation, for the more arduous one of experiment." To test the worth of his invention, "it was not absolutely necessary to understand minutely the theory, but rather to observe the effects. It is by these that the existence of a cause is ascertained, and by these also its utility is demonstrated. The phenomena are often learned from observation long before we arrive at the theory which connects and explains them."[11] Perkins thus appeals to common sense, understood as a tried-and-true faculty that judges by experience and keeps its eyes on the sort of things "that daily happen in the world," eschewing fantasy, theory and speculation. But the common sense invoked so rhetorically by Perkins is common in the additional, and less reassuring, respect that it is a sense that people possess above and beyond their five native senses. Perhaps common sense pronounces in favor of the Perkins tractor only because we are so responsive to one another that we expect to feel as our fellows feel, or as we are given to believe they do. With so many reports of relief and so many authorities vouching for the instrument—with the thing enveloped in clouds of publicity—is it not possible that those trying it would form a sort of imaginary judgment of its powers? According to Johnson, the novel that shuns "incredibilities" and confines itself to the terrain of common sense, with characters like us inhabiting a world like our own, seems to appeal specially to the modern imagination. Like responds to like. Perkins rhetoric also appealed to the imagination, making the tractor seem all the more wondrous because it proved itself even to skeptics and men of science, and priming prospective users for bodily sensations that others like them were reported as having.

The first investigations of what is now known as the placebo effect were reactions to Mesmerism and the Perkins tractor. Disturbed by the proliferation of not exactly secret societies devoted to Mesmerism and by the rampant success of such a questionable theory-practice, Louis XVI in 1784 appointed a commission of eminent scientists, chaired by the skeptical Benjamin Franklin, to look into animal magnetism. In a masterful experiment the investigators found that subjects who did not know that a certain tree or cup of water had been magnetized responded as they would to any tree or cup of water, and conversely that subjects who believed things magnetized that had not been magnetized at all acted as if in the presence of a mysterious force. The investigators concluded that animal magnetism had no physical existence. As reported to the Academy of Sciences in September of that year, the commission of inquiry

> discovered we could influence [blindfolded subjects] ourselves so that their answers were the same, whether they had been magnetized or not. This means we were dealing now with the power of the imagination. . . . We succeeded in manipulating the imagination. Without being touched or signaled, the subjects who thought themselves magnetized felt pain, felt heat, a very great heat. In some cases, we provoked convulsions and what is known as crises. The subjects' imagination could be brought to the point of the loss of speech. It allowed us to produce all the so-called effects of magnetism, even the calming down of convulsions.[12]

But as elegant as these experiments were, indeed precisely because they were elegant, they could not reproduce the frenzy that surrounded Mesmerism and gave it power—the impassioned public chatter, the heightened, almost sexual interest the subject seemed to elicit, all of which combined to charge the experience of those who came within its range. An experiment artfully conducted in Benjamin Franklin's gardens could not possibly capture the communalism of the Mesmer phenomenon. The mechanism of Mesmerism—the means by which its theorized power flowed from one body to another—was not so much the maestro's wand and theatrical props as the atmosphere surrounding the craze itself, an excitement impossible to simulate in a controlled experiment. Though trees and water could not be magnetized, the followers of Mesmerism might still be magnetized to one another.

Similarly, while neither the flow of energy known as animal magnetism nor the universal harmony theorized by the mesmerists existed, ideas themselves flowed across time and space during the Enlightenment, generating excitement as they went—among them the conception and design of the glass harmonica that provided Mesmer's unearthly sound-effects. When the maestro adopted the instrument, he could not have known that the man who perfected and named it would preside over the exposure of Mesmerism itself as a shared delusion. In a letter to Beccaria of 1762, Benjamin Franklin laid out the construction of the harmonica in detail, so that "you, or any of your friends, may be enabled to imitate it, if you incline to do so, without being at the expence and trouble of the many experiments I have made in endeavouring to bring it to its present perfection."[13] Built into the music accompanying mesmeric ritual, then, was the same practice of methodical experimentation that would be deployed against Mesmerism itself. Franklin saw a prototype of the glass harmonica made by a member of the Royal Society, who in turn got it from a certain Mr. Puckeridge in Ireland. The passage of the concept from Ireland to Franklin in London, from Franklin in turn to Beccaria in Turin, and from whatever source to Mesmer in Paris, where he met Franklin in 1779, illustrates in its own way the propagation across distance not only of information but the sort of potential energy bound up in it.

Fifteen years after the inquest into Mesmerism, when skeptical investigators tested the Perkins tractor against a wooden facsimile in the General Hospital of Bath, the report of their experiment—a historic paper by John Haygarth entitled *Of the Imagination as a Cause and as a Cure of Disorders of the Body*—cited the precedent of the Franklin commission.

> It need not be remarked, how completely the trial [of the Perkins tractor] illustrates the nature of this popular illusion, which has so wonderfully pre-

vailed, and spread so rapidly; it resembles, in a striking manner, that of Animal Magnetism, which merited the attention of FRANKLIN, when ambassador from America, and of other philosophers at Paris.[14]

That this was written in Bath, where a surgeon acted as Perkins' sales agent, suggests something of the scope of a craze born in New England. (Indeed, Franklin's son became an officer of the Perkins Institute in London.) We can well imagine the circulating reports and roused expectations that carried the Perkins phenomenon across an ocean—all the way to Denmark, in fact—and made a virtual cult object of something that could be manufactured in a barn.

In an advertisement in the *Times* of London on December 21, 1798, Perkins cited in his own favor comments from the *Medical and Chirurgical Review* of September and October of that year that bring out both the interest excited by the tractor and the importance of testimonials in spreading its reputation:

> No inconsiderable degree of importance attaches to the subject. The testimonies in support of the Tractors are at least honourable, if not decisive. Many of them come from persons not of the medical profession, but who nevertheless seem competent on the present occasion: for the subject is for the most part an appeal to the common senses. Others of the facts are vouched for by members of the medical profession of high and distinguished character. Unquestionably, if there be no error or misconception of facts, the importance of the discovery is great indeed. Where the proofs are so many and so strong, it would be unreasonable to question them on any other grounds than cautious and fair experiments.

If this passage can be considered as a challenge by Perkins to his doubters and detractors, then Haygarth took it up.

Days later, on 7 January 1799, reports Haygarth, five patients suffering from chronic rheumatism were treated with wooden devices indistinguishable to the eye from the Perkins tractor, and all but one "assured us that their pain was relieved," thus demonstrating that the actual composition of the instrument had nothing to do with the relief derived from it. But how exactly was treatment administered? "The wooden Tractors were drawn over the skin so as to touch it in the *slightest* manner," but this in itself was not probably enough to induce an experience of relief. It seems the investigators also made sure to tell the patients about the relief that others had already experienced; for Haygarth advises anyone seeking to replicate his results to conduct the trial "with due solemnity. During the process, the wonderful cures which this remedy is said to have performed ought to be particularly related."[15] In this sense, the power source of the Perkins instrument, like that of Mesmerism, was other people. If patients who had never heard of the instrument and knew nothing of its mystique had their skin brushed with a rod, be it of wood or metal, it seems improbable that they would react like Haygarth's five whose expectations were deliberately inflamed, still less like people swept up in a common transatlantic enthusiasm. One of Haygarth's fellow investigators, working at the Bristol Infirmary, reported being beset by so many volunteers for the Perkins treatment that he had to hurry his ministrations; "yet such effects were produced as were almost incredible."[16] That crowd might serve as a symbol of the sort of electricity that flowed through the Perkins phenomenon, whether or not the tractor was in touch with electricity itself. Members of the community of believers were inspired by one another. In one case the Bristol investigator informed a patient (no doubt with an inaudible laugh) that "I had an instrument in my pocket which had been very serviceable to many in his state,"[17] before proceeding to treat him with a sham tractor. In effect, the patient was invited to feel what others like him felt; and so too, we can safely assume, were the original subjects of Haygarth's experiment—among the earliest to employ an inert simulacrum of a questioned treatment.

Both the Perkins tractor with its homespun simplicity and Mesmerism with its exotic flamboyance came wrapped in words. Neither could have caught on and become a movement otherwise. Both Mesmer and Perkins enlisted the printed word, with its power to be everywhere, to advertise their access to an also universal power. Both movements no doubt throve on rumor and wonder as well as the sort of testimonials that Mesmer is said to have carried in his pocket and Benjamin Perkins cited at length in a pamphlet. Had Johnson lived to witness the two phenomena, he would probably have marked them down as instances of "the dangerous prevalence of imagination," although imagination aroused not by solitude, as in *Rasselas*, but by a communal fever. Johnson may have been more right than he knew to remark on our attraction to stories of others like ourselves undergoing credible experiences. As the kindred phenomena of Mesmerism and Perkins-ism suggest, even bodily experiences may mirror those of others, or those imputed to others. Indeed, the mysterious universal powers the two therapies claim to tap, powers that flow through us whether we know it or not, resemble nothing so much as the power of example itself, which as Johnson observes "is so great as to . . . produce effects almost without the intervention of the will."[18]

The unreliability of our experience, along with our susceptibility to influence, was dramatically demonstrated almost simultaneously with Haygarth's experiment a few miles away in Bristol, where Thomas Beddoes and Humphry Davy were investigating the therapeutic potential of nitrous oxide. Knowing that sensational reports about the gas and its effects were in circulation and could very well color the experience of study subjects, Davy "devised a novel method of randomised control: some volunteers were first given a bag of common air to check that their response was due to the gas rather than their imagination." But not only study subjects were suggestible. In the evenings members of the experimenters' circle inhaled the gas for fun and proceeded to behave in a wild, but not too wild manner that may have been induced by the gas but was certainly also both licensed and expected by those looking on. It was in part because of the theatrics connected with the inhalation of the gas in such a charged social setting that the use of nitrous oxide in surgery, where it would eventually prove itself so impressively, was not even considered by the investigators.[19]

⇌

Although in the case of nitrous oxide the social character of experience made for noise that obscured a medical signal, it is an important principle in its own right. Recently, when subjects were asked to inhale a gas described as a suspected environmental toxin but which in fact was common air, they proceeded to report at a high rate exactly the same unpleasant symptoms supposedly reported by others exposed to the mystery gas.[20] In a reverse placebo experience, they felt what they presumed others did. The predisposition to do just this constitutes one of the channels of the placebo effect itself.

Thus, in a recent study of "placebo analgesia through social observation," a confederate of the experimenters reported a reduction in pain associated with a certain green light. Observing subjects then reported a "robust" reduction in their own pain in association with the same light, even though the association was entirely fictitious. So feelingly did they enter into the experience, or the supposed experience, of a present other that their own pain-experience changed accordingly. Elsewhere subjects derive a reduction in pain from a placebo without seeing others enjoying the same presumed benefit, but simply by hearing about it. In one study of treatments for irritable bowel syndrome, subjects treated with a placebo were told "the substance you've just been given has been shown to powerfully reduce pain in some people in previous studies,"[21] at once an equivocation and a strong cue. In a study that did away with the equivocation, some IBS patients were given a simulacrum explicitly described as "placebo pills, something like sugar pills," but said to have been "shown in rigorous clinical testing to produce significant mind-body self-healing processes."[22] Strongly invited in this way to experience a therapeutic effect that others enjoyed, and that seemed validated by the authority of science (instead of being pulled as it were from the pocket of the experimenter as in one of the Bristol trials of the Perkins tractor), a significantly higher percentage of the placebo group than the no-treatment group reported improvement. If this experiment supports the possibility of open placebos—a matter I will look into in due course—it also suggests that the tendency to feel what others feel is strong enough to moot the knowledge that one's treatment is "something like" a sugar pill. In encouraging patients to model their very experience on the reported experiences of others, the designers of the study, knowingly or not, followed a precedent now two hundred years old.

By the same token, however, the principle that we tend to blend our experiences with others' as imagined or reported has implications outside the laboratory. Consider a study of the analgesic effect of branded and unbranded aspirin, which found not only that the former outperformed the latter in relieving headaches but that branded placebo outperformed unbranded placebo. Why such confidence in a label? We might suppose that study subjects had taken the brand of aspirin in the past and found it worked—acquired their confidence by experience. However, "branding was effective for nonusers as well as users, which may suggest that the particular brand tested has a general reputation for efficacy."[23] (But what is it that endows a brand of pill with a reputation for efficacy, such that someone who has never taken it implicitly trusts it? It can only be advertising, which suffuses our world like another universal medium, and which found its modern shape in the third quarter of the 18th century in England.[24]) If I believe in the efficacy of a brand of pill I do not take, this is tantamount to saying I assume it will reduce my pain because it seems to reduce that of others—with "others" in this case going beyond those I see (as in the green-light experiment) or even those I am told of (as in the open placebo experiment) to include an indefinite multitude. My very sensations may be mediated by strangers.[25] But at this point we return to the original investigation of the placebo effect.

Chapter Six

To Feel What Others Feel
Mind and Body

Were early investigators of the placebo effect guilty of mind-body dualism?

In contrast to clinical trials where the placebo effect represents nothing but a confounding factor, recent research—including studies of the vicarious reduction of pain or the use of explicit placebos to treat pain—envisions it as something interesting and potentially fruitful in its own right. But before writing off the skeptical estimate of the placebo effect as devaluing and reductive, we do well to reflect on the research two centuries ago from which it arose, including the work of John Haygarth (1740-1827).

Among the earliest studies of the placebo effect, as opposed to a treatment that unwittingly exploited it, was the simple yet ingenious trial conducted by Haygarth, a physician, in Bath in 1799. At the time, as we know, the fashion for extracting the body's pains by the use of wand-like instruments known as tractors had spread from the United States to England. Produced and patented by the American Elisha Perkins and sold at the impressive price of five guineas (twenty-five dollars on the other side of the Atlantic), these homely articles of brass and iron found enthusiasts in the medical world itself, as the cited notice in the *Medical Chirurgical Review* clearly implies.[1] It was Haygarth, remembered mainly for his effort to check the spread of smallpox by means of mass inoculation, strict quarantine, and sanitary regulations,[2] who first tested the alleged benefits of "Tractoration" in something like a controlled experiment.

Having fashioned a wooden object indistinguishable to the eye from the Perkins tractor, Haygarth proceeded to treat five patients at the General Hospital suffering from chronic rheumatism by passing the instrument over their skin, all the while letting them understand that it was the celebrated tractor.[3] While the experiment would not meet today's standards of informed consent or methodological rigor, it did make a bold and far-reaching point, one expounded in a paper presented by Haygarth to the Literary and Philosophical

Society of Bath and then published in 1800. Inasmuch as the Perkins tractor was puffed in pamphlets citing what are advertised as hard facts, it can be said that Haygarth answered Perkins in kind by circulating the findings of his experiment in printed form.

Investigating the imagination "as a Cause and as a Cure of Disorders of the Body," Haygarth records the procedure and results of his experiment before citing the testimony of other investigators. Each of his subjects, he notes, had been ill for months before being treated with false tractors on 7 January 1799.

> All the five patients, except one, assured us that their pain was relieved, and three much benefited by the first application of this remedy. One felt his knee warmer, and he could walk much better, as he shewed us with great satisfaction. One was easier for nine hours, and till he went to bed, when the pain returned. One had a tingling sensation for two hours. . . . Such is the wonderful force of the Imagination!
>
> Next day, January 8th, the true metallick Tractors of PERKINS were employed exactly in like manner, and with similar effects. All the patients were in some measure, but not more relieved by the second application, except one, who received no benefit from the former operation. . . . It need not be remarked, how completely the trial illustrates the nature of this popu-lar illusion, which has so wonderfully prevailed, and spread so rapidly. . . If any person would repeat these experiments, it should be done with due solemnity. During the process, the wonderful cures which this remedy is said to have performed ought to be particularly related. . . . The whole effect undoubtedly depends upon the impression which can be made upon the patient's Imagination.[4]

Contrary to the pamphlet's title, in this case imagination certainly did not cure a malady, for if the same patients were treated for the same symptoms on consecutive days, first with a look-alike and then with the Perkins trac-tor, the relief credited to the wooden devices cannot have been very lasting. Nevertheless, the appeal of the Perkins cult is understandable. At a time when patients were routinely bled and purged in the effort to expel illness quite literally bodily, the Perkins tractor, so it seemed, removed what ailed the body without breaching it at all—and without doing harm[5]—by means of some force unknown, but apparently on the order of electricity or (though the younger Perkins indignantly denied it) magnetism. According to Benjamin Perkins, George Washington himself was a proselyte. It is also recorded that in an episode of fever at the age of 57, Washington was bled some 2.5 quarts and purged repeatedly, a treatment that could have killed him.[6] Well might he have preferred the Perkins tractor.

But the thing was worthless, as Haygarth demonstrated. Trained at the University of Edinburgh under the renowned professor of chemistry William Cullen, who was a friend of Hume's (and a user of placebos in his medical practice),[7] Haygarth had skepticism in his intellectual bloodline and called on it in his experiment on the Perkins tractor, or rather on those who believed in it. In the process he identified very clearly what came to be known as the placebo effect: the ascription of medical properties to inert treatments as a result of the belief invested in them. The placebo effect was responsible for much of the good done even by reputable doctors in Haygarth's time, and remains to this day a potent variable to be accounted for in medical and pharmaceutical research. Though he did not name it and was not the first to take note of it, Haygarth was among the first to employ a placebo experimentally. The now-contested view of the placebo effect as a sort of canard reflects the Enlightenment critique of superstition which in turn informed Haygarth's thinking about the Perkins tractor.[8]

That imagination wields power over the body and its ills was no discovery of the eighteenth century, however. In Chaucer's Miller's Tale, when John the carpenter is fed a tall tale about an impending flood of Biblical magnitude, he quakes with fear, thinking he can actually see it coming. "Men may die of imagination," the Miller exclaims satirically. But if the imagination can harm, can it also heal? In a discourse on The Force of Imagination, Burton, the anatomist of melancholy, cites an entire litany of unusual cures. "As some are . . . molested by phantasy," he records,

> so some again, by fancy alone, and a good conceit, are as easily recovered. We see commonly the toothache, gout, falling sickness, biting of a mad dog, and many such maladies cured by spells, words, characters, and charms. . . . All this world knows there is no virtue in such charms or cures, but a strong conceit and opinion alone. . . . The like we may say of our magical effects, superstitious cures, and such as are done by mountebanks and wizards. "As by wicked incredulity many men are hurt . . . we find in our experience, by the same means many are relieved." An empiric oftentimes, and a silly chirurgeon, doth more strange cures than a rational physician. Nymannus gives

a reason, because the patient puts his confidence in him, which Avicenna prefers before art, precepts, and all remedies whatsoever.

Burton concludes with a maxim already known to us: "'Tis opinion alone ... that makes or mars physicians, and he doth the best cures, according to Hippocrates, in whom most trust."[9]

However close this estimate of imagination's empire over the body may appear at first glance to Haygarth's, Burton belongs to an order that still looks to Avicenna and Hippocrates as medical authorities, a quasi-medieval cosmos held together not by laws or law-like principles but by all manner of resemblances, sympathies, and associations. A few lines after the cited passage Burton instances the case of a carcass that "bleed[s] when the murderer is brought before it."[10] Just as strangely, he seems to assume that the remedies of wizards and vendors actually do work, one and all, provided only that the sufferer believes in them. Confidence—belief—trumps "art, precepts, and all remedies whatever." Though Burton was surely aware of the pejorative connotations then often attaching to the word "opinion," he maintains—or at least his authorities do—that in medicine opinion is all. He shows no interest in testing the cures wrought either by the rational physician or the charlatan.

The foremost doctor in England in Haygarth's time—Erasmus Darwin—reported a case in which a woman caught stealing firewood called down a curse of cold on her apprehender, and so vivid was his fear of it that he kept to his bed for upwards of twenty years, "till at length he died": a story worked up in Wordsworth's "Tale of Goody Blake and Harry Gill," where the curse has actually taken effect.[11] Experimental investigation of the placebo was born of the Enlightenment's interest in the imagination as the source of unusual effects, as in Darwin's report, along with its concern with the verification of knowledge. For unlike Burton, Enlightenment thinkers *were* interested in testing cures. Close in spirit to Haygarth's test of the Perkins tractor, and cited by Haygarth, is the inquest into Mesmerism conducted by Benjamin Franklin and others, including Lavoisier, at the behest of the King of France in 1784. Subjecting themselves to animal magnetism, the investigators experienced nothing, as befit philosophers, in high contrast to all those thrown into paroxysms by the occult fluid. Haygarth's research into the power of the imagination situates itself in this lineage. With talk of electricity and animal magnetism in the air, and the Perkins movement attracting followers with its own sort of magnetism, it is believable that patients treated with sham tractors would feel the tingling they imagined others felt.

Haygarth's investigation of the placebo effect represents an impresive exercise of Enlightenment intellect. Committed as he was to the public exercise of reason, a stickler for evidence, Haygarth would have been immediately suspicious of a cure-all like the Perkins tractor whose composition was patented, that is, kept secret.[12] One who presented his findings to the Literary and Philosophical Society of Bath could not be expected to believe in something explicitly guarded from investigation. He could not be expected to believe in a wand at all. A fellow investigator quoted in his pamphlet refers to simulacra of the Perkins tractor as "wonder-working pieces of wood,"[13] a description that puts them in the same category as relics, icons, and other medieval superstitions, yet what was the tractor itself but a wonder-working piece of metal? The attempt to trace the placebo effect to its origin in the human mind constitutes in itself an Enlightenment mode of reasoning, as when Hume opens his *Treatise of Human Nature* by looking into "the Origin of our Ideas." The understanding of the imagination as both a source of delusion and a potential cause of demonstrable effects belongs, too, to the Enlightenment.

⊟

Haygarth's pamphlet begins, "That faculty of the mind which is denominated the Imagination, has been the subject of two very elegant compositions in the English language, in prose and poetry, by ADDISON and AKENSIDE. It has not wholly escaped the notice of medical writers, but merits their farther investigation."[14] Significantly for our purposes, not only does Addison devote a Spectator paper to the pleasures of the imagination, he judges these pleasures beneficial to health.

> We might here add, that the Pleasures of the Fancy are more conducive to Health than those of the Understanding, which are worked out by Dint of Thinking, and attended with too violent a Labour of the Brain. Delightful Scenes, whether in Nature, Painting, or Poetry, have a kindly Influence on the Body, as well as the Mind, and not only serve to clear and brighten the Imagination, but are able to disperse Grief and Melancholy, and to set the Animal Spirits in pleasing and agreeable Motion. For this reason Sir *Francis Bacon*, in his Essay upon Health, has not thought it improper to prescribe to his Reader a Poem or a Prospect . . .[15]

Addison's idea that the pleasures of the imagination are beneficial for mind and body—note the pairing—looks back to the theory that it does us good to be transported from time to time from our cares, as when Boccaccio's brigade of nobles flees a plague-stricken city to indulge in the delights of storytelling. It is interesting that Haygarth should have paid tribute to Addison in a paper devoted to exposing a treatment whose efficacy was purely imaginary.

The satiric estimate of the imagination to which Haygarth leans is less in the spirit of Addison than of Johnson.[16] A chapter of Johnson's *Rasselas* warns famously of "The dangerous prevalence of imagination," an admonition inspired by the royal party's encounter with an astronomer who has somehow managed to convince himself that he determines the course of the sun and rules the seasons. The man is a study in the "prevalence" of imagination in that fantasy overrides his reason, but while it is hard to believe there are many such on the face of the earth, clearly he is held up by Johnson as an example (if an extreme one) of an ordinary human failing, one that "prevails" in the sense of being of general extent or of common occurrence.[17] "There is no man," says the wise Imlac, commenting on the astronomer, "whose imagination does not sometimes predominate over his reason. . . . No man will be found in whose mind airy notions do not sometimes tyrannise."[18] Imagination in this sense is not the poet's distinction but humanity's common lot, not a glory but a trap. Johnson once remarked over dinner, "Were it not for imagination, Sir, a man would be as happy in the arms of a Chambermaid as of a Duchess. But such is the adventitious charm of fancy, that we find men who have violated the best principles of society, and ruined their fame and their fortune, that they might possess a woman of rank."[19]

In the same year as *Rasselas* (1759) appeared Adam Smith's *Theory of Moral Sentiments*. Those who think of Smith as an apologist for commercial society will be surprised to discover the satiric view he takes of some of its motive forces and highest prizes. According to Smith—a sometime table companion of Johnson and also, for what it's worth, a close friend of Haygarth's mentor, William Cullen[20]—the reason we sweat and strain to get ahead in the world is simply to become like the rich who enthrall our imagination.

> From whence, then, arises that emulation which runs through all the different ranks of men, and what are the advantages which we propose by that great purpose of human life which we call bettering our condition? To be observed, to be attended to, to be taken notice of with sympathy, complacency, and approbation, are all the advantages which we can propose to

derive from it. . . . The rich man glories in his riches, because he feels that they naturally draw upon him the attention of the world.[21]

But how can a mere craving for admiration account for the restless activity of commercial society? How can such a paltry cause account for such a mighty effect? Only because imagination magnifies the object of desire—the privileged life—into something greatly to be wished.

> When we consider the condition of the great, in those delusive colours in which the imagination is apt to paint it, it seems to be almost the abstract idea of a perfect and happy state. It is the very state which, in all our waking dreams and idle reveries, we had sketched out to ourselves as the final object of our desires.[22]

Thus, the spirit of striving that animates commercial society springs from a deception of the imagination, though Smith is prepared to rate this a benign deception. After explaining that imagination confuses the splendor of the rich with "the order, the regular and harmonious movement of the system, the machine or oeconomy by means of which it is produced," he comments, "It is well that nature imposes upon us in this manner. It is this deception which rouses and keeps in continual motion the industry of mankind."[23] No longer the capricious tyrant it is in Burton, imagination now serves to strengthen civil order itself.

Somewhat like the difficult word "opinion" in Shakespeare's time, which had strong connotations of instability and epistemological license but was not limited to these senses, "imagination" in the age of Smith might or might not carry connotations of fallacy. Smith himself does not always regard the imagination as an instrument of delusion (whether happy or otherwise) or as divorced from reality. The opening sentences of the *Theory of Moral Sentiments* look into the case of a person who enters by imagination into the suffering of another to the point of trembling and shuddering, a case that raises the possibility that imagination can act on the body. As Smith sees it, we "enter as it were into [the] body" of the one in suffering "and become in some measure the same person with him."[24] The possibility that mind can influence body is taken up by the dedicatee of Haygarth's pamphlet, his colleague William Falconer, whose celebrated essay of 1788 on *The Influence of the Passions upon Disorders of the Body* explores the proposition that the mind can both cause and cure disorders, as in the case of a man long afflicted with gout who upon being sentenced to death but pardoned at the last minute, found his limbs restored to "activity and strength, whereas before that event their use was nearly lost."[25] Haygarth intended his experiment to show that the Perkins tractor induces the credulous to imagine themselves cured, or at least relieved. According to Falconer, however, the condemned man really was cured. "This person . . . lived many years totally free from the gout." So too, in the case of intermittent fever,

> It is well known that numerous cures . . . have been performed by medicines of little, or even of no medical efficacy whatever in themselves, which effect could proceed only from the opinion the patient entertained of their powers; as a proof of which we find that the certainty of the cure has almost always depended on the degree of the patient's confidence in the success of the remedy.[26]

In other words, in certain cases of this disorder treatments without medical merit do seem to work (though one wonders exactly where Falconer came by his knowledge of each patient's degree of confidence). Not only does Haygarth cite this essay whose findings regarding the medical import of the imagination seem so unlike his own—so nearly Burtonian—but evidently he places his pamphlet in the same discourse. It seems his skeptical understanding of the imagination had room for his colleague's argument that imagination may in fact engender physical effects, just as Falconer's appreciation of imagination's power did not keep him from endorsing Haygarth's skeptical experiment "entirely."[27] Indeed, the two men collaborated to fashion the false tractors.

Some who argue today that the placebo effect cannot be reduced to a fallacy born of a sugar pill also dispute mind-body dualism, stating or implying that it is built into modern Western medicine. Recognition of the possible influence of mind on body is built into Haygarth's pamphlet—as paradoxical as this might seem—in the form of citations of Falconer. For the fact is that while duly skeptical of quack cures and faith-healing, Falconer's essay on *The Influence of the Passions upon Disorders of the Body* finds that emotions like fear and hope contribute to the prevention, course, and cure of many a disease by virtue of their profound if obscure connection to the life of the body. While it is hard to believe that Falconer would credit Tractoration as a cure for rheumatism, he does recognize the possibility in other cases of cures produced by worthless treatments. So too does a live toad worn around the neck staunch a bloody nose—not because of any medical magic inhabit-ing the toad but because the creature acts so powerfully on the mind. "It is not improbable that the sentiments of aversion, dread, and horror, impressed by such an odious contact, may act as a powerful sedative, and of course be serviceable in the disease, by diminishing the force of the circulation."[28] The curious double view of the imagination taken in Haygarth's historic pa-per—as a source of delusion on the one hand and actual effects on the other, anticipating the paradox of the placebo response as both fallacy and potent mechanism—is also to be found in Falconer, though differently inflected.

Not only, however, does Haygarth cite Falconer's research into the imagi-nation in exposing the Perkins tractor, but he himself states both in his title and his text that imagination can produce real effects. He might even concur with Wordsworth's claim in the Preface to the *Lyrical Ballads*, published the same year as his exposé of Perkins, that "the power of the human imagi-nation is sufficient to produce such changes even in our physical nature as might almost appear miraculous,"[29] as both he (Haygarth) and Falconer cite at length James Lind's account of the cure of scurvy by means of a spurious medicine purportedly sent by the Prince of Orange during the siege of Breda in 1625.[30] Toward the end of Haygarth's pamphlet cases are cited that "prove (what should be well understood) that the Imagination can cause, as well as cure, diseases of the body." Therefore, while Haygarth considers the Perkins tractor a demonstrable sham—his title brands it "fictitious"—he does not picture the imagination as an entity cut off from the body and its health. Far from insisting on some kind of mind-body dualism, he concludes that "It is extremely fortunate, when the most powerful remedies of both body and mind unite their power to alleviate or remove a disease."[31] The Perkins trac-tor, it seems, free-rides on a legitimate medical phenomenon, and is all the more disreputable for that reason.

The common claim that the prevailing understanding or misunderstand-ing of the placebo effect derives from the "Cartesian dualism" of modern medicine[32] is misleading. Descartes himself, in a work proverbially said to sever mind and body, states explicitly that the mind is "dependent upon the humors and condition of the organs of the body," and looks forward to an improvement of medical knowledge enabling the cure of "an infinity of mala-dies of body as well as of mind."[33] Underlying the work of Falconer and Haygarth, too, was a tradition of medical thinking whose members deemed it

> their business as physicians to set forth the circumstances, extent and means
> of control of the interaction between mind and body. They believed . . . in

keeping with an unbroken tradition originating in ancient Greek medicine, that the mind could be influenced corporeally, that is, by means of drugs, diet, climate and other factors acting primarily on the body. Clinical observation had convinced them that the cause of a bodily disturbance could lie in the mind. . . On the other hand, it had also convinced them that the causes of some mental disorders lay in the body, and they drew the obvious inference regarding treatment.[34]

This tradition was not abolished but modified and extended in the Enlightenment.

A novelist who, like Haygarth, received medical training in Edinburgh depicts a world where bodies act on minds and feelings have physical effects—where indigestion alters the brain and the belief that the ragout one has just eaten was cat induces "convulsive agonies."[35] "I am equally distressed in body and mind," says the author of the first letter in Smollett's *Humphry Clinker*, writing to his doctor.[36] David Hartley, founder of the theory of association, concluded that "there ought to be a great reciprocal influence between the mind and the alimentary duct."[37] Reform circles of Haygarth's day took up the principle of association and wove it into doctrines which "by eliminating the mind-body distinction, provided a scientific basis for the . . . belief that social and moral behavior could be changed by bodily discipline," as in a well-designed prison or indeed hospital.[38] (By the same token, however, a reformer like Mary Wollstonecraft could argue that girls should have more physical freedom, because "dependence of body naturally produces dependence of mind.")[39] Among the Hartleyans was the hospital reformer Thomas Percival, whose treatise on *Medical Ethics* (1803) is sometimes cited in the placebo literature. However, Haygarth's "Tractoration" experiments offer no evidence of an intention to use a hospital as a laboratory to mold citizens, impose discipline on the lower orders, correct their thinking, or advance any other heavy-handed project to reform his fellow human beings. The experiments are an exercise in enlightened skepticism, not disciplinary ambition.

☒

Especially given the polemical commonplace that modern medicine rests on mind-body dualism, it is worth noting that the mind-body distinction was by no means fixed either in the discourse of Haygarth and Falconer or the philosophy of the Enlightenment. The physician Locke, whose influence presides over the Enlightenment, complicates the distinction between mind and body even as he affirms it, stating that "By Pleasure and Pain I must be understood to mean of Body or Mind, as they are commonly distinguished; though in truth they be only different constitutions of the Mind, sometimes occasioned by disorder in the Body, sometimes by Thoughts of the Mind."[40] Arguably, the task of doing justice to so mixed a phenomenon as the placebo effect calls for a qualified skepticism in the tradition of the Enlightenment, and in particular of Haygarth and Falconer—the former of whom debunked a treatment that played on the imagination but confessed that the same faculty could be a medically potent force, while the latter had great medical respect for the imagination but seconded Haygarth's plan to expose the Perkins tractor as an imposition on just that. And to the names of Haygarth and Falconer we might add Benjamin Franklin.

It was in Franklin's gardens near Paris, in an experiment of elegant design, that Mesmerism was conclusively exposed as an artifact of suggestion. Franklin's skepticism extended to medicine itself, the one merit of Mesmerism in his estimation being that its false promises deterred people from ruining their health by taking drugs. "There are in every great city a number of persons who are never in health, because they are fond of medicines and always taking them, whereby they derange the natural functions and hurt their constitutions. If these people can be persuaded to forebear their drugs in expectation of being cured by only the physician's finger or an iron rod pointing at them, they may possibly find good effects, though they mistake the cause."[41] That expectation unlocks the placebo effect is no recent discovery.

According to Poor Richard, "He's the best physician that knows the worthlessness of the most medicines." Poor Richard also maintains that temperance itself is health and that it results from the good government of the body by the mind, just as, conversely, moderation of diet subdues the passions and keeps the mind in good order.

> Wouldst thou enjoy a long Life, a healthy Body, and a vigorous Mind, and be acquainted also with the wonderful Works of God? Labour in the first place to bring thy Appetite into Subjection to Reason. . . .
>
> A sober Diet makes a Man die without Pain; it maintains the Senses in Vigour; it mitigates the Violence of Passions and Affections.
>
> It preserves the Memory, it helps the Understanding, it allays the Heat of Lust; it brings a Man to a Consideration of his latter End.[42]

A healthy body and a vigorous mind seem to go together; the mind works on the body, the body on the mind. "Mind-body dualism" seems like an inaccurate description of this arrangement.

While his inquest into Mesmerism helped establish understanding of the placebo effect as a sort of trick perpetrated by the mind on the body, Franklin would surely have been interested in the Falconer-Haygarth investigation of the medical connections between the two.[43]

Chapter Seven

The Invalidism of Mr. Woodhouse

By dint of having his fantasy indulged by everyone around him, the man's imaginary ailment has become real.

The inverse of a vigorous mind in a healthy body would be a frail mind in a frail body. Such is the state of Mr. Woodhouse as we encounter him in the brilliant opening chapter of Jane Austen's *Emma*, set immediately after the marriage, and therefore the departure from the household, of Emma's governess Miss Taylor. Completely unequal to his patriarchal position and said to be "without activity of mind or body," the enfeebled Mr. Woodhouse stands as a sad exhibit of the interdependence of the mental and the physical. The text suggests that he sank to the condition in which we find him not as a result of the natural course of any disorder but because he abandoned himself to an imaginary helplessness to which others deferred, wisely or not. If companionship can itself be a medicine—alleviating gloom and distress, for example, as a certain tradition of common sense maintained—so too, it seems, can the aid of others complicate our troubles. Mr. Woodhouse is a study in socially constructed invalidism.[1]

The void at the moral center of Emma's world, Mr. Woodhouse is identified as "having been a valetudinarian all his life," which makes him (according to the Oxford English Dictionary) "A person in weak health, esp. one who is constantly concerned with his own ailments; an invalid." In the more provocative terms of Falconer's treatise on mind-body medicine, Mr. Woodhouse suffers from or cultivates something like hysteria, albeit without such physical symptoms as "paroxysms." The same irony that makes him a figure of imperious debility gives him a malady more commonly imputed to women.[2] Of hysteria Falconer writes that "nothing contributes to aggravate it more than indolence and vacancy of mind"—Mr. Woodhouse's vocation. Concerned lest people accede to the hysteric's fictions and affectations, Falconer cautions that

too great solicitude to avoid every thing likely to give uneasiness, especially
if such solicitude be very apparent, is likely to do as much mischief as service. Nothing so much enhances the apprehension of danger, or so causes
those on whose account the care is taken, to believe that the hazard is greater than it really is, and such circumstances frequently recurring, keep them
perpetually in a state of painful irritation, which in reality constitutes the
disorder. It would be much better to inure such persons gradually to the
common occurrences of life.[3]

Seven years have not been enough to inure Mr. Woodhouse to his elder
daughter Isabella's marriage, which he persists in regarding as a misfortune
to her because it feels like one to him. Naturally he views the marriage of
Miss Taylor earlier that day in the same light. Even a walk on a mild moonlit
night seems to Mr. Woodhouse, in his hysteria, a "shocking" event, and to
venture half a mile to visit the newlyweds is out of the question.

> "My dear, how am I to get so far? Randalls is such a distance. I could
> not walk half so far."
> "No, papa, nobody thought of your walking. We must go in the car-
> riage to be sure."

This little exchange well illustrates the sort of care and coddling Mr. Woodhouse's hysteria receives—at least from Emma, his principal prop, for whom
his incapacity means her own freedom.

In view of her father's infantilism, the wonderful first sentence of the
novel takes on new meaning:

> Emma Woodhouse, handsome, clever, and rich, with a comfortable home
> and happy disposition, seemed to unite some of the best blessings of exis-
> tence; and had lived nearly twenty-one years in the world with very little to
> distress or vex her.

The inference is inescapable: having a father who demands to be tended like
a child does not disagree with Emma Woodhouse. And perhaps it is because
this arrangement suits her that Emma is shown pampering the nominal pa-
triarch, playing along with his hysteria in violation of Falconer's sensible
advice. At one point she even denies he is living in a fantasy world, thereby
reinforcing the most fundamental delusion of all. Tellingly misconstruing a

remark by Mr. Knightley as concerning him, Mr. Woodhouse confesses, "I am afraid I am sometimes very fanciful and troublesome," as if begging to be contradicted. Emma instantly reassures him that it is not so. "My dearest papa! You do not think I could mean *you*, or suppose Mr. Knightley to mean *you*." That Mr. Woodhouse is immediately corrected by Emma after his all-too-accurate admission shows just what a cobweb of fictions his life has become, and how the complicity of others, in particular Emma, keeps the delicate structure from being swept away. "His spirits required support."

Following a superb exchange between Emma and Mr. Knightley about whether she engineered the match between Mr. Weston and Miss Taylor or it came about by itself, Mr. Woodhouse, "understanding but in part," makes a simple-minded comment. If he had understood the exchange, it might have set him thinking about whether his own strange condition came about by itself or as the result of social complicity. That Mr. Woodhouse has been "a valetudinarian all his life" means he is not only wealthy enough to make inactivity his occupation but important enough, at least locally, to command deference despite himself. He is, we are told, "a much older man in ways than in years; and though everywhere beloved for the friendliness of his heart and his amiable temper, his talents could not have recommended him at any time," which suggests that his pretense or affectation of inability has become all too real through lack of doing anything and through the cultivation of his weaknesses by those who love him.[4] Second nature seems to have become nature itself. The most he may be capable of is treating others as if they were as delicate as himself.[5]

Instead of the remedies recommended by Falconer, such as "to inure [the hysteric] to the common occurrences of life," Emma abets Mr. Woodhouse's fantasy that he is too frail for life itself. If to feel as others feel, or seem to feel, is a channel of the placebo effect, Mr. Woodhouse is forever unable "to suppose that other people could feel differently from himself," and so projects onto them his own cultivated inability to support life's occurrences. (Hence his offer of an egg to Mrs. Bates in chapter 3: "Mrs. Bates, let me propose your venturing on one of these eggs. An egg boiled very soft is not unwholesome. . . . You need not be afraid—they are very small, you see—one of our small eggs will not hurt you.") The supposed frailty of others then serves to justify his own, although his, by virtue of long pretending, has become real enough. If a placebo effect lies somewhere on the continuum between the suppositious and the real, in the case of Mr. Woodhouse an affected delicacy has become genuine incompetence.

In chapter three of *Emma* we hear of certain newfangled schools in which "young ladies for enormous pay might be screwed out of health and into vanity." Without the help of any system or theory but certainly with the aid of others, Mr. Woodhouse has sacrificed his wellbeing to a fantastic obsession with health itself.

⊟

If the magic of words is such that being able to name their condition gives some patients a sense of control of it (a transaction sometimes called the Rumpelstiltskin effect), a medical label can also have the reverse effect of contributing to a sense of illness. A study published in 1978 found that after a screened group was diagnosed with hypertension, their annual rate of absenteeism increased by 80%, whether or not the condition was actually being treated.[6] The diagnosis, the label, evidently underwrote the workers' understanding of themselves as ill even though their illness had no symptoms and the normal range of blood pressure is a matter of ambiguity. The label Mr. Woodhouse wears as a valetudinarian, a man in permanent ill health, makes for a sort of moral absenteeism, in the form of an abdication of his patriarchal responsibilities.

Some medical labels may even give stage directions for the performance of illness. A recent discussion of the nocebo phenomenon—the induction of adverse effects by expectation—points out that given groups not only have their own classifications of disease but in effect their own scripts for acting it out. A nosology of disease

> is also a *sickness repertoire*, available for performance. . . . Knowledge that symptoms such as fainting exist provides a role or script available to be performed. In addition, nosologies may be *licenses* (insofar as they certify the cultural legitimacy of the condition) or *prescriptions* (insofar as they define expected sequences of occurrences). However, nocebo acts need not be—and most appear not to be—deliberate, voluntary, or fully conscious.[7]

Perhaps the understanding of valetudinarianism in his local world gives Mr. Woodhouse a "legitimate" script to perform (not that he could get away with it if he were a servant rather than a man of leisure). It is understood that a certain sort of person, called a valetudinarian, is in weak health; constantly concerned with his own health; an invalid. This definition does not exactly state that the invalidism in question may have been self-caused.[8]

Hysteria as explained by Falconer lacks legitimacy both because it is such an artificial construction and because it is bound up with "indolence and vacancy" as well as "selfish and mean ideas and sentiments,"[9] all of which Mr. Woodhouse exhibits unmistakably. It seems that the available sickness repertoire has given Mr. Woodhouse a role to play that is all at once congenial to his temperament, congenial to Emma (in that it nullifies his authority), entitled to the deference of others, and finally debilitating. Hysteria as instanced in Mr. Woodhouse is a nocebo phenomenon.

What of the qualification that nocebo acts need not be conscious and deliberate? That Mr. Woodhouse never intended to lose himself in a pathological fantasy goes without saying. But never would he have been able to do so without continual babying, especially by Emma herself; and though Emma certainly enjoys and abuses the freedom she acquires as a result of his abdicated authority, she is not so ill-meaning as to undermine him with cynical intent. Others, too, defer to Mr. Woodhouse without ill designs. Even the sensible Mr. Knightley, who contests Mr. Woodhouse's hysterical interpretations as far as possible, can only say so much.

The last words of the opening chapter belong to Mr. Knightley, the irony of their implications ringing out like the after-tone of a bell. Let Mr. Elton choose his own wife, he tells Emma. "Depend on it, a man of six or seven-and-twenty can take care of himself." Yet Mr. Woodhouse cannot take care of himself and apparently never could and never will. The novel ends with Mr. Woodhouse in the same state of incapacity as at the beginning, as if the changes that come to others—the events of his world—had simply passed him by. Where in the first chapter he is said to hate "change of every kind" and marriage in particular as the "origin of change," in the novel's last paragraphs he proves himself unchanged by dreading the impending marriage of Emma and Mr. Knightley. What snaps him out of it is not "any sudden illumination" of his mind or "any wonderful change of his nervous system," we are told. It happens that some turkeys are stolen from the nearby poultry-houses, a crime so alarming to Mr. Woodhouse that he is relieved to put himself under the protection of his soon-to-be son-in-law. One dread simply displaces another.

Being "leveled with the rest of the world,"[10] Mr. Woodhouse is ineligible for sudden illuminations and wonderful changes—for miraculous cures—although in fact his case does not lend itself to cure at all. Because he hates change of every kind he must also hate the thought of the only thing that could help him, a reformation of his own habits. Over time, through indolence and the deference of others, he has gotten so locked into these practices that any departure from them is now outside the realm of the possible.

⛎

Our own time has seen any number of hitherto unknown ailments identified and popularized—ADHD, chronic fatigue syndrome, premenstrual syndrome, and social phobia, for example, all of which may be responsive to placebo treatments. What's wrong with the expansion of the list of medical or psychiatric disorders and the marketing of illness that goes along with it? A well-informed critic of the trend toward the medicalization of human life can only say that the classification of so many deviations from the medical average as illnesses deprives the world of "diversity," in effect making it a less colorful place.[11] But Mr. Woodhouse is revealed as a colorful man precisely in that he has renounced maturity in order to play the child and devote himself to his fantasies, and so too may an abundance of new disorders make the world a more variegated place merely by their presence. Let a hundred flowers of illness bloom. Far from representing a police action by a medical community bent on suppressing diversity, the installation of novel entities such as chronic fatigue syndrome, adult ADHD, and even multiple personality disorder in the American Psychiatric Association's *Diagnostic and Statistical Manual* often comes about at the urging of groups seeking to have their claim to difference—their diversity—officially recognized.[12]

A staunch believer in chronic fatigue syndrome has argued that a child with any of its symptoms should have complete rest, inasmuch as "the stress of leading a normal life and keeping up with their peers can exacerbate the condition," which as others have noted is a prescription for invalidism.[13] And at this point we return to the case of Mr. Woodhouse, for whom incapacity became a way of life. The trouble with the ever-expanding list of ailments and the tidal trend toward the medicalization of life is not that they make things bland and uniform but that the creation and popularization of disorders which may or may not have a medical basis not only provide templates of debility to model ailments on (thereby encouraging their spread) but pose the same risk to which Mr. Woodhouse succumbed: the risk of the patient growing into the complaint.

Chapter Eight

The Transfusion of Life
Tolstoy's "Master and Man"

A case that defies the social model of the placebo effect in every detail.

W hen Anatole Kuragin proposes to elope with Natasha in *War and Peace*, she feels powerless to resist, maintaining "I have no will."[1] So recklessly and yet scrupulously does Natasha abandon herself to the fantasy that romantic love takes over one's entire being that she imagines herself, for the moment at least, as Kuragin's "slave." Upon learning that he is actually married, she attempts suicide, albeit ambiguously, by swallowing "a little" arsenic, after which she sinks into a condition Tolstoy names grief, marked by such symptoms as loss of appetite, loss of sleep, and low spirits. The account of her treatment constitutes a remarkably explicit description of medical treatment as a social proceeding.

> Doctors came to see her singly and in consultation, talked much in French, German, and Latin, blamed one another, and prescribed a great variety of medicines for all the diseases known to them, but the simple idea never occurred to any of them that they could not know the disease Natasha was suffering from, as no disease suffered by a live man can be known, for every living person has his own peculiarities, and always has his own peculiar, personal, novel, complicated disease, unknown to medicine—not a disease of the lungs, liver, skin, heart, nerves, and so on mentioned in medical books, but a disease consisting of one of the innumerable combinations of the maladies of those organs.

But how can doctors fail to perceive something as blindingly obvious as their own futility? Because as vain as their ministrations are, they fill a need.

> Their usefulness did not depend on making the patient swallow substances for the most part harmful . . . but they were useful, necessary, and indispens-

able because they satisfied a mental need of the invalid and of those who loved her—and that is why there are, and always will be, pseudo-healers, wise women, homeopaths, and allopaths. They satisfied that eternal human need for hope of relief, for sympathy, and that something should be done, which is felt by those who are suffering. . . . The doctors were of use to Natasha . . . assuring her that [her illness] would soon pass if only the coachman went to the chemist's in the Arbat and got a powder and some pills in a pretty box for a ruble and seventy kopecks, and if she took those powders in boiled water at intervals of precisely two hours, neither more nor less.[2]

The doctors, then, set social wheels in motion—there are orders to follow, errands to run, rituals to fulfill—and it is all this make-work centered on the patient, not the actual composition of the powders and pills, that does some good, albeit more for the household than the patient. The patient, we are told, does find all the concern lavished on her "pleasant," which corresponds to the literal meaning of "placebo": "I shall please." The placebo effect gives Natasha some comfort while her condition remits in its own time, not as a result of the charade of medical ministrations.

In a late Tolstoy story one party comes to another's aid in a way that entails no show of activity and far transcends the placebo effect. The doctors in *War and Peace* do not know what Natasha is suffering from—they don't even know she attempted suicide—but the master in "Master and Man" (1895) knows his servant is freezing because both are caught in the same blizzard. There is nothing inferential about his comprehension of the servant's plight, no room for supposition or theory. That the servant actually wants to live is indeed open to question, but that he will die unless somehow kept warm is not. As if Tolstoy had constructed a case specifically excluding the placebo effect, the master transfuses into the servant not morale but physical warmth. As Natasha heals not because of medicine but nature ("youth prevailed," says Tolstoy), as disease in "The Death of Ivan Ilych" derides the efforts of medicine,[3] so does the blizzard in "Master and Man" sweep away all defenses against its power except the elemental heat of the body itself.

According to a social understanding of the placebo effect, the very company of others may do us good, inspiring us with hope and reminding us that we belong to something larger than ourselves—the social body. "Humans are social animals, even in our grief," writes a physician, reflecting on the placebo effect. "'Misery loves company.'"[4] Misery does love company in "Master and Man," albeit not in the sense that it wants others to be wretched or even brightens in their presence, as with traditional social therapies for melancholy. Here company in suffering kindles the discovery of love.

Obsessed with the thought of making a killing on a piece of land, the confident Vasili Andreevich Brekhunov, a merchant and church warden whose name suggests "Liar," sets out for a certain forest with his servant Nikita one winter afternoon despite threatening weather. A blizzard sets in, and as the markers identified by the master dissolve one after another like mirages in a desert of snow, the two become lost and end up doubling back on their own track. The sort of running in rings that does some good when it is undertaken for Natasha's benefit, does no good at all in a storm at night. Master abandons servant and sets out into the darkness (at once naturalistic and symbolic, like the fable itself), only to be abandoned in turn by his horse and flooded with "such terror that he did not believe in the reality of what was happening to him."

> "Queen of Heaven! Holy Father Nicholas, teacher of temperance!" he thought, recalling the service of the day before . . . He began to pray to that same Nicholas the Wonder-Worker to save him, promising him a thanksgiving service and some candles. But he clearly and indubitably realized that the icon, its frame, the candles, the priest, and the thanksgiving service could do nothing for him here and that there was no and could be no connexion between those candles and services and his present disastrous plight.[5]

The equivalent of the communion administered to the dying Ivan Ilych as if it could somehow console or prepare him, the props and trappings of Orthodoxy as here portrayed have no value beyond providing a kind of evanescent sensation of relief, if that; at best, they are a psychological placebo. Only when he comes upon the horse and the sledge where Nikita is hunkered down does the master's terror pass, and when it does so his contempt and indifference toward the servant pass as well. It is the fact of another life, encountered in this stark setting as if for the first time like a natural miracle, it is this fact that awakens the master "clearly and indubitably" to the reality of love, and rescues him from both desperation and despair as prayer could not.

So fortifying is his reunion with Nikita that "if he felt any fear it was lest the dreadful terror should return that he had experienced when on the horse and especially when he was left alone in the snow-drift." When Nikita says he is dying,

> Vasili Andreevich stood silent and motionless for half a minute. Then suddenly, with the same resolution with which he used to strike hands when making a good purchase, he took a step back and turning up his sleeves began raking the snow off Nikita and out of the sledge. Having done this he hurriedly undid his girdle, opened out his fur coat, and having pushed Nikita down, lay down on top of him, covering him not only with his fur coat but with the whole of his body, which glowed with warmth.

The following morning the master is discovered frozen to death, but the servant is alive.

The transfer of life from master to servant unfolds much less predictably than a placebo event governed largely by expectation. The servant certainly cannot expect to receive life from his master. Indeed, nothing about the case seems to conform to the model of a placebo transaction in which, say, one party feels something presumed or reported to be experienced by others. ("This pill has been shown to help others . . .") Not only does the narrative concentrate more on the donor than the receiver, but the master's warmth passes into the body of the servant in a physical and elemental manner worlds apart from an operation dependent on suggestion or report. Just as significantly, the master—the source of the warmth—is himself changed profoundly by this transfusion of life. When the covered Nikita begins to stir, the masters says,

> "There, and you say you are dying! Lie still and get warm, that's our way . . ." He stopped speaking and only gulped down the risings in his throat. "Seems I was badly frightened and have gone quite weak," he thought. But this weakness was not only not unpleasant, but gave him a peculiar joy such as he had never felt before. . . .

> He did not think of his legs or of his hands but only of how to warm the peasant who was lying under him. . . . He could not bring himself to leave Nikita and disturb even for a moment the joyous condition he was in. He no longer felt any kind of terror.

Whereas earlier he prayed to the Wonder-Worker to save him, he himself now saves Nikita, and in so doing discovers love—not its semblance but its reality. As opposed to one who receives a transient feeling of comfort or consolation from a church ritual (the same ceremony enacted by others in the same way), the master experiences a cessation of terror that remains with him as he lies motionless for hours on end, right up to his death, which he understands to be death.

Before going about warming his servant, the master is shown standing "silent and motionless for half a minute." What goes through his head Tolstoy doesn't say, but it seems unlikely that he is making suppositions or inferences about Nikita's state. Why would he? He knows exactly what Nikita is suffering without having to go through mental exercises because they are both subject to the same storm and, of course, to mortality itself. So too, unlike Ivan Ilych, whose wife prevails on him to take communion because "it often helps,"[6] the master goes about warming the servant with no expectation of the joy that awaits him; when it arrives it is experienced as something never foreseen or imagined, let alone felt. If the placebo effect is sometimes envisioned as a message sent by the mind to the body, the master, concerned only with warming "the peasant who was lying under him," seems to think with his body. "Tolstoy's contemporaries noted that he alone appreciated the way the body shapes the mind."[7]

The transaction between master and man exhibits none of the trappings of expectation and suggestion, of inferred experience and questionable reality, that surround the placebo effect. They have all been stripped away.

⊟

As a social phenomenon, the placebo effect takes place in a region of high ambiguity. In the account of Natasha's illness, doctors appear at once useless and useful, and by joking with the patient "regardless of her grief-stricken face"[8] they act out a charade that she herself seems to find not unpleasant. (Similarly, though portrayed as headstrong fools, their conclusion that "the malady was chiefly mental" is not foolish at all.) In the opening chapter of *Anna Karenina* the charming but dishonest Stiva, a social creature par excellence, has been caught in adultery by his wife and tries to maneuver his way out by putting on his habitual smile, now become an ambiguous grin of guilty innocence. As if pantomiming "placebo" ("I will please"), Stiva evidently hopes this assumed expression will be so pleasing, so disarming and irresistible, that all will be forgiven; maybe, by a kind of reciprocation of sentiments, Dolly will even smile back. His hope is disappointed.

The transfusion of the master's warmth into Nikita has nothing in common with this sort of wished-for transfusion of sentiments, or indeed with the experience of second-hand effects. It is not a transaction like the passage of a smile from one person to another and therefore defies the model of the placebo effect as a social proceeding. The master's joy does not enter the servant. Quite unlike one who experiences a benefit (perhaps subjective, perhaps not) as a result of a social influence, Nikita does not even think of himself as helped. Nothing better illustrates the refutation of the placebo effect in "Master and Man" than his displeasure at having survived. "When he realized that he was still in this world he was sorry rather than glad." It seems the master has saved the life of one for whom life is a matter of indifference at best.

Though we naturally want to romanticize a tale where a master dies for his servant and finds his life by losing it, the text itself opposes us, and nowhere more stubbornly than in its insistence that the servant was ready for death when the master revived him. The fatalistic Nikita thinks little of life, after all, and though dutiful would probably not have done for the master what the master did for him—not because he was unwilling to relinquish life but because it would have meant taking action. Saved by means of a deed he did not ask or wish for, he is especially sorry to find "that the toes on both his feet were frozen." For twenty years he lives on, it is reported, never once grateful to the man who literally laid down his life for him. It is as if he had been cheated of his rightful death. When he finally does die he is "sincerely glad that he was . . . now really passing from this life of which he was weary into that other life which every year and every hour grew clearer and more desirable to him."

A placebo should not harm. Company and good fellowship, traditional therapies for melancholy, do not harm. Though worthless except insofar as it deterred users from harmful treatments, the Perkins tractor was innocuous; it was simply waved over the body. Communion, says the wife of Ivan Ilych, often helps and "can't do any harm." By contrast with that ritual repudiated by the heterodox Tolstoy, the transfusion of life into Nikita takes place without the mediation of words and symbols; but no less significantly, the prolongation of life received as a result, unasked-for and at the cost of several toes, is unwelcome to Nikita. To the servant's way of thinking he was indeed harmed, although—in one more instance of resignation—he accepts his injury without resentment.

⊟

Tolstoy's was not the only imagination summoned by the image of the physical transfusion of life. In *The Brothers Karamazov*, Ivan confesses to Alyosha,

> I could never understand how one can love one's neighbours. It's just one's neighbours, to my mind, that one can't love, though one might love those at a distance. I once read somewhere of John the Merciful, a saint, that when a hungry, frozen beggar came to him, he took him into his bed, held him in his arms, and began breathing into his mouth, which was putrid and loathsome from some awful disease. I am convinced that he did that from 'self-laceration,' from the self-laceration of falsity, for the sake of the charity imposed by duty, as a penance laid on him.[9]

Flaubert told the similar story of St. Julian the Hospitaller warming a leper with his body, a rendition Tolstoy found skillful but cold and unmoving, lacking the sincerity of great art. In Tolstoy's working of the motif, which has been transposed from distant to modern times, the beneficiary of charity is not a leper and there is no suggestion that the act of charity was imposed by duty or is in any way played up. Clearly, the author wishes to portray an action at once remarkable, admirable, yet believable, and to that end he sets his tale in the here and now and removes the trappings of the miraculous, much as he has also excluded the placebo effect.

That Tolstoy thus rules out the placebo effect, with its social transfusion of experiences, does not mean that he rules out social transfusion as such. Indeed, according to the theory of art he composed around the same time as "Master and Man," the distinguishing quality of true art is precisely that it "infects" us with another's sentiments. Under its influence we experience "that simple feeling . . . of being infected by the feelings of another, which makes us rejoice over another's joy, grieve over another's grief, merge our souls with another's."[10] The placebo experience, by which we feel what we imagine others feel because we take part in the same rituals or are swept up in the same trends and movements—this, it turns out, is but a degraded parody of the way truth is communicated. Art transfuses life into our frozen souls.

Chapter Nine

The Pollyanna Principle

The presumption against deception should not be discarded lightly.

When Natasha in *War and Peace* descends into listlessness and despondency for causes known to the reader but not her doctors, her symptoms tally quite well with the diagnostic criteria of depression as first laid out in the third edition of the *Diagnostic and Statistical Manual* (1980).[1] Over the decades since its diagnostic criteria were codified in the DSM, the incidence of depression seems to have sky-rocketed. What to do about it?

Although the placebo effect may seem like an underpowered weapon to use against a problem of such alleged gravity and magnitude, that is just what a recent paper provocatively advises. Contending that depression is best treated with psychotherapy designed to instill "positive illusions," and that medical bodies therefore need to rethink their position on the permissibility of deception, the author recommends for the depressed a healthy dose of a placebo called the Pollyanna Principle. The ability to hold false but encouraging ideas about themselves and the world makes for mental health (so it is argued), and is associated in the well with "an increased capacity to care for others and an enhanced aptitude for creative and productive work." Under the benevolent rule of the Pollyanna Principle, it seems, we take on a resemblance to the inhabitants of a utopia like William Morris's Nowhere, who are indeed portrayed as identically empathetic, creative, productive, and happy.

Given that antidepressants, for all their popularity, generally not only do not work appreciably better than placebos but have nontrivial side effects, the author—a faculty member in a School of Politics, International Studies and Philosophy—argues that a better way to treat depression must be found; and given that deception or self-deception is said to be an ingredient of mental health itself, if not its foundation, the author furthermore argues that the ban on deception in medical practice ought to be lifted. The treatment recommended for depression is an intensive, protracted, and deliberately deceptive course of cognitive behavioral therapy.

This is a form of therapy which encourages patients to identify negative thought patterns and subsequent behaviour and to consider whether such thought patterns and responses are useful or helpful to them. Over a long period of time (many months), patients are persuaded to adopt thoughts which are more "realistic" (but which are in fact moderately positive) and which induce different behavioural responses. . . . Individuals who successfully complete such forms of treatment end up endorsing positive illusions about themselves. . . . These psychological therapies involve a more prolonged form of deception than placebos; any deception about the efficacy of prescribed sugar pills pales when contrasted with the promotion of highly personal deep-seated illusions about oneself that are induced in the successful treatment of a patient with depression.

According to the author, "Medical bodies need to accept that a spoonful of deception may be fundamentally (and unavoidably) therapeutic,"[2] although a course of deception extending over "many months" seems less like a spoonful than a steady infusion. And while "unavoidably therapeutic" seems to mean that successful treatment cannot be achieved without deception, in point of fact much of what is called depression is "likely to abate over time without intervention."[3] A deceptive treatment lasting months on end may succeed not because of the deception per se but because it allows time for a transient condition to pass of its own accord.

Although depressed patients are to be induced to think "realistic" thoughts, one of the author's principal sources maintains that realism is part and parcel of depression itself.[4] Patients must therefore be cured of realism without being tipped that this is in fact taking place. Once persuaded to adopt the illusions cultivated by most people (for "normal people possess unrealistically positive views" of themselves and the world),[5] their treatment itself is successful, their case closed; they have learned to see as others do. In the 1960s Herbert Marcuse, soon to be a mentor of the student revolt, derided the utopia of a consumer society that both mandates and mass produces a false happiness. "The Happy Consciousness—the belief that the real is rational and the system delivers the goods—reflects the new conformism which is a facet of technological rationality translated into social behavior."[6] Yesterday's protest has become today's proposal. The prospect of a society investing its power, wealth and credit in the production of happiness by the methodical deception of citizens one by one—in effect, its own deception—is as strange as it is chilling.

"It is the job of physicians," we are told, to "restore positive illusions" in depressed patients by psychotherapy. But physicians are not psychotherapists and are not going to commit themselves to working for "many months" to adjust a patient's thinking. By the same token, though, actual psychotherapists, not being medical doctors, are perfectly free exploit the placebo effect—and do. While medical regulations like the AMA Code of Ethics (cited by the author) restrict the use of placebos, there is no corresponding provision in, say, the American Psychological Association's code of ethics, which says nothing about placebos. The argument that depression requires the cultivation of encouraging half-truths, and that therefore medical bodies should re-write their codes of ethics, misses the point that psychotherapy is not constrained by medical codes of ethics. If it were, it would not have been portrayed some years ago in Frank and Frank's *Persuasion and Healing*, a landmark of the literature, as an institution that plays on the placebo effect and builds morale by fostering beliefs that are healthy and "satisfying" but not necessarily true,[7] which is approximately what the Pollyanna paper urges right now. As I will argue, the freedom to exploit the placebo effect in psychotherapy—a habitat uniquely adapted to it—has had something to do with the surging popularity of that institution just when the physician's right to use placebos came sharply into question.

Also in *Persuasion and Healing*, it is noted that practitioners of cognitive-behavioral therapy—the mode of therapy recommended in the Pollyanna proposal—"explicitly instruct new patients about the therapeutic task in such a way as to strengthen their expectations. . . . The therapist tells the patient at length about the power of the treatment method, pointing out that it has been successful with comparable patients and all but promising similar results for him too."[8] So did Haygarth inflame the expectations of subjects by telling them of the cures performed by the Perkins tractor, so does the doctor ordering vitamin injections cite their benefits for other patients, and so even now (as I will document) do experimenters with open placebos take care to remind subjects of the treatment's success in other cases like theirs. With striking similarity, all play on the placebo effect's social sources. In the case of the Pollyanna proposal, however, the spoonful of truth in the claim that treatments that work for some will work for others conceals an expansive right to lie.

While white lies can certainly be justified as indispensable to social life, the lying recommended in the Pollyanna paper is too prolonged and methodical to be written off as incidental fibbing. The principle that the truth of psychological counsel doesn't really matter, or even matters inversely, has risks unrecognized in light-and-easy defenses of the Pollyanna Principle. The sort of abuses inseparable from paternalism were documented by Sissela Bok early in the era of informed consent,[9] and the proposed deception of the depressed for their own benefit constitutes one more form of exactly that— paternalism. That the deception is carried out by a therapist does not exempt it from objection; there is nothing about psychotherapy that releases it from the moral considerations that apply to other human activities. Indeed, as a *program* of deception, at once systematic, intensive, and conducted with an elaborate show of professional benevolence, the proposed enterprise goes far beyond common lying. The authority it would accrue makes its risks that much more serious. The principal source cited in the Pollyanna paper in *defense* of positive illusions concedes that

> a falsely positive sense of accomplishment may lead people to pursue careers and interests for which they are ill-suited. Faith in one's capacity to master situations may lead people to persevere at tasks that may, in fact, be uncontrollable; knowing when to abandon a task may be as important as knowing when to pursue it. Unrealistic optimism may lead people to ignore legitimate risks in their environments and to fail to take measures to offset those risks. . . . Faith in the inherent goodness of one's beliefs and actions may lead a person to trample on the rights and values of others.[10]

Notably, in their influential defense of positive illusions the authors of these words do not claim that happy people tilt "moderately" toward such illusions; on the contrary, we are told that "far from being balanced between the positive and the negative, the perception of self that most [happy] individuals hold is heavily weighted toward the positive end of the scale."[11] Perhaps it is just because the illusions they have in mind are so pronounced and potent that the authors do *not* suggest providing them to those in need, which would be playing with fire.

It is also notable that the master, Brekhunov, in "Master and Man" is shown at the beginning of the tale brimming with positive illusions. He is so good at self-deceit that he actually convinces himself he is not stealing from Nikita. On the day he sets out to make his purchase, using over two thousand rubles of church money in his possession, "he was even more pleased than usual with . . . all that he did."[12]

The proposition that human life would be poorer without the solace of fantasy is not a new one. According to Erasmus's Lady Folly, humanity is kept happy by ignorance, imbecility, and forgetfulness; especially blessed is the species of folly that "comes about whenever some genial aberration of mind frees it from anxiety and worry while at the same time imbuing it with the many fragrances of pleasure."[13] Folly is nature's antidepressant. Nowhere does Lady Folly suggest that delusions be administered to the population by certified experts, if only because they would then lose their genius, their inspiration.

Asks Bacon in his essay "Of Truth," "Doth any man doubt, that if there were taken out of men's minds vain opinions, flattering hopes, false valuations, imaginations as one would, and the like, but it would leave the minds of a number of men poor shrunken things, full of melancholy and indisposition, and unpleasing to themselves?" Without the consolation of fiction, it seems, we are vulnerable to depression. But note that while Bacon performs the mental experiment of removing "vain opinions" to see what remains, he does not prescribe delusions for those whose store may be low, nor does he question the supremacy of truth. Our love of the lie, though "natural," is also "corrupt," he contends. It was Pilate who said in jest, "What is truth?"[14]

Far from prescribing deceptions and beguilements, traditional thinking about melancholy emphasized the therapeutic value of the sort of counsel that is so plainly true that one wants to call it a truism. Thus the "comfortable speeches" and "consolatory speeches" instanced by Burton in *The Anatomy of Melancholy* as examples of good advice point out that things are not as bad as they may seem, that others suffer too, that not everything can "answer our expectation," that matters could be worse, that "if naught else, time will wear [sorrow] out; custom will ease it; oblivion is a common medicine for all losses, injuries, griefs, and detriments whatever." Though Burton well knows such commonplaces may leave us cold—"Most men will here except: Trivial consolations, ordinary speeches, and known persuasions in this behalf will be of small force"—he esteems the traditional consolations all the same. "Yet sure I think they cannot . . . but do some good, and comfort and ease a little."[15] Sooner will he serve up a proverb like the healing power of time than a therapeutic dram of deception.

A corollary of the feeling for the mutability of things that deeply informs literature (think of Hamlet's "But two months dead" or the overnight reversal of fortune in "Master and Man"), the principle that time wears sorrow out has much truth. It is just because much of what is classified as depression is "likely to abate over time without intervention" that a course of therapy lasting months may seem effective when the effective agent is time itself, just as any number of cases of improvement credited to placebo treatments ever since Beecher may actually have arisen spontaneously. Traditional thinking about melancholy or depression is structured by the distinction between sadness arising from the events of life itself—and therefore liable to subside with the flow of time—and excessive, habitual sadness. This distinction has fallen into neglect, as in the Pollyanna proposal; hence, perhaps, the alarmingly high incidence of depression cited to justify a modest proposal to deceive millions of people for their own good. "In the USA alone," we are told, "diagnostic rates [of depression] are estimated at around 10% of the adult population per annum."

At this point we are confronted with the paradox of depression's popularity.

Chapter Ten

Medicine Marketed: Two Episodes

Antidepressants are enveloped in advertising and folklore, and so was the
Perkins tractor.

According to a common estimate, then, fully 10% of American adults
qualify as depressed. Desperate problems call for determined measures;
thus the proposal to beguile the depressed population one by one into a state
of normative happiness. Upon examination, though, the 10% figure proves
to be inflated, so that the crisis that serves to justify a national program of
deceptive therapy falls to the ground, and the justification with it. If you
check the source for the figure cited in the Pollyanna proposal, you find that
it is explicitly intended to show how and why such exaggerations come into
being and pass as coin of the realm. "In fact, the extraordinarily large number of people who allegedly suffer from categorical mental disorders is a
product of symptom-based measures [that is, simplistic diagnostic checklists]
that inevitably overestimate the number of people who have some untreated
mental illness," as Allan Horwitz demonstrates in an important study of the
way conditions like depression have come to be defined and diagnosed.[1] The
traditional understanding of depression or melancholy as uncaused, habitual
or inordinate sadness (as in Burton) has been abandoned, with the result
that much normal sadness is now classified as a pathological condition. One
reason for antidepressants' reputation for efficacy, despite their questionable
performance in clinical trials, may be that many who take them do not really
suffer from a disorder in the first place.

It seems the claim that 10% of the adult population is depressed actually
means that something like that number take antidepressants. But even that
figure understates the market for these drugs. "About 10 percent of Americans over age six now take antidepressants," reports the former editor of the
New England Journal of Medicine, not to alert the public to an epidemic
of alarming magnitude but to deplore the gross inflation of the estimated
incidence of depression.[2] It seems more probable that such estimates reflect

the aggressive marketing of treatments for depression, in the context of the medicalization of life in general, than an authentic epidemic. "For a generation accustomed to translating human conditions into medical terms," writes Edward Shorter, "unhappiness becomes 'depression,' a curable disorder."[3] Concerned, like others, with the medical import of our social connections and disconnections, Shorter traces this unhappiness to the instability of the bonds that support wellbeing—in particular the fragility of intimate bonds in a postmodern society—arguing, however, that people are not actually becoming more ill but simply "more *sensitive* to the symptoms they already have."[4] For those suffering more from the impression of an ailment than an ailment per se a placebo-like medication is made to order—all the more when, like an antidepressant, it comes wrapped in the mystique of science, the glare of advertising, and the haze of popular mythology.

Along with its staggering dimensions, another aspect of the depression epidemic stands out: the unimpressive showing, when tested against placebos, of the drugs so successfully marketed to treat the condition. In 2002 a review article in *JAMA* found that the response-rate to placebo in published trials of antidepressants is "often substantial," adding intriguingly that it has "increased significantly in recent years."[5] If a pill seems to many a more direct and efficient way of tackling mental illness than the more laborious path of psychotherapy,[6] perhaps that is because it offers a more economical dosage of placebo than a course of treatment extending over months. The drug makers too have run into the wall of the placebo effect. Mimicking response-rates to the drugs themselves, "the placebo response [in antidepressant trials] is unpredictable and seemingly unmanageable, and costs drug companies hundreds of millions of dollars in failed trials and delayed or shelved compounds."[7] The story goes that Merck had high hopes for a certain promising antidepressant, only to discover in advanced clinical trials that it was no more effective than a pill containing no medication at all.

A common factor—the mass marketing of drugs for depression—contributes to (a) the exaggerated scale of the depression epidemic and (b) the strikingly high rates of placebo response in trials of potential agents. As for (a), it is enough to note that within a few years after the FDA approved direct-to-consumer advertising of medications in 1997, spending on antidepressants more than doubled, and within a decade the drugs were among the most popular medications in the United States. If advertising in general beckons us to fulfill our desires, antidepressant ads hold out the radical and exceptionally seductive possibility of removing what stands in the way of fulfilling our desires—and doing so in company with countless others, virtually as members of a movement. Even before the approval of DTC ads, antidepressants had generated their own folklore. Indeed, so great were public and professional enthusiasm for the new wonder drugs that it is estimated that between 1987, when Prozac was introduced, and 2002 "almost one in four people in the United States was started on a Prozac-type drug."[8] Again, it is far more likely that these drugs, by popular repute able to make patients "better than well," created their own market[9] than that they brought to light an ignored population of astonishing magnitude. Prozac rode the "utopian wave" it generated[10]—the consumer side of the therapeutic fervor that mandated that every child in America be screened for depression[11] regardless of the crudity of the screening instrument, the predictably spurious estimates of prevalence that would ensue, and the dubious wisdom of prescribing antidepressant medication for children.

An already-cited study conducted three decades ago found that aspirin "supported by extensive advertising" significantly outperformed its generic counterpart.[12] In the case of antidepressants, the very fact of advertising sent a potent message, namely that the time for depression to step out of the shadows had arrived, so that those who suffered from it or believed they did, or whose symptoms could be matched with this ambiguous disorder, or who were not happy and interpreted their state as a clinical condition, could now find hope and encouragement—the stuff of the placebo effect (b). Mass marketing changed the game, not only introducing the public to certain compounds but paradoxically popularizing depression itself, portraying it as a socially legitimate condition and thereby selling morale (more clinically, "mood") along with the allegedly specific benefits of the drugs themselves. As the *JAMA* review article noted in 2002, "In recent years, as effective treatments for depression have become more widely available *and socially acceptable*, there has been a marked increase in the proportion of the population receiving treatment for depression."[13] There is a clear connection between numbers treated and the marketing of pharmaceutical remedies; perhaps a link also exists between the efficacy of the treatments and their social acceptability.

Labels can be potent, after all.[14] The label of "valetudinarian" seems to give Mr. Woodhouse an identity and a vocation, though the effect in this case is enfeebling. The diagnosis of "irritable bowel syndrome" makes many people patients whose abdominal symptoms are no different from many more who do not seek medical assistance. In the study of branded and unbranded aspirin, not only did the former outscore the latter, but branded placebo outscored its unbranded equivalent. However, in the same study the users of the branded pills also reported significantly more headaches, as if a greater imputed power to manage headaches allowed for more of them. This sort of paradox—greater incidence side by side with experienced relief, all mixed in with the placebo effect—may also be at work in the antidepressant phenomenon, with labels in this case belonging both to the pill and the ailment itself. The marketing of antidepressants raises the incidence of depression, while the labeling of the condition as an ailment (rather than a failing) may encourage and console. If, as some say, merely by acquiring a name for their disorder patients can gain a sense of control over it,[15] this might be all the more likely in the case of a disorder as blurry, conceptually ambiguous, and ominous as depression. The labeling of depression as a reputable as well as treatable condition, and this by manufacturers whose power was now being mobilized for the benefit of the affected, meant that it was no longer necessary to suffer in silence—to be depressed about being depressed.[16] Morale and expectation were mobilized so effectively that the placebo effect in trials of potential agents rose even as the numbers diagnosed as depressed did as well. A socially constructed epidemic[17] brought with it a socially fueled placebo factor.

With depression becoming a growth market, millions became theoretically depressed less as a result of a severe condition than of the trend toward diagnostic inflation, the elasticity of descriptors, the fashion for "brain" instead of "mind" explanations, and the availability of much-heralded treatments. Moreover, as prescriptions shot up and antidepressants acquired a mythology and a popular literature (most notably the best-seller *Listening to Prozac* [1993]), the campaign against depression began to resemble a movement, as if it had something in common with another enterprise that brought people out from the shadows and into the light: the civil rights movement. Enlisting hope and drawing on the prestige of science and the ardor of advocates as well as the craft of Madison Avenue, the pharmaceutical campaign on behalf of the depressed tapped deeply into non-pharmaceutical sources. Rising numbers themselves can be inspirational, with the diagnosed now forming a sort of brotherhood. Misery loves company.[18] Antidepressants may or may not target the cause of depression, but the message of antidepressant advertising—"You are not alone, millions of others share your problem"[19]— certainly targets the placebo effect, inviting the receiver of the message to experience the same uplift as so many others. "The placebo effect reminds us that we are not alone."[20] Those consumers of antidepressants not really suffering from severe depression in the first place, but from something less grave and more transitory misinterpreted as a pathological condition—and the evidence suggests they were and are many—would of course be a natural constituency for a placebo response. The steep increase in diagnosed depression, coinciding with all-out marketing of drugs for depression, coinciding with the popularization of depression and the formation of something resembling a depression movement,[21] coinciding with a recorded increase in the placebo response—at some point, all this becomes an exercise in mimesis.

A critic of the antidepressant myth has drawn an analogy between these drugs, imagined as aiding the flow of signals in the brain, and Mesmerism, imagined as aiding the flow of an elemental fluid, animal magnetism.[22] Let me propose a parallel between antidepressants and another 18th-century therapy that inspired research into the placebo effect (and was likened by its critics to Mesmerism): the Perkins tractor. This is not to say the analogy is complete—for antidepressants may offer some benefit above and beyond placebo, particularly in cases of severe depression, while on the other hand they are not as harmless as the Perkins treatment—but that both mobilize a placebo effect with a strong social component.

1. Two Marketing Phenomena

Powered in the first instance by advertising, but also by lore, mythology, and public enthusiasm, the market for antidepressants represents a triumph of salesmanship. So did the Perkins tractor. Indeed, only if invested with a potent mystique could such a homely article have attracted so many takers in the first place. Its simplicity may have fed that mystique, evoking not only straightforward Yankee virtue but the lightning rod and, through it, an elemental force of Nature. The inventor's son dedicated himself to defending and promoting the instrument—documenting its successes in print (the medium of publicity) and extending its empire across the Atlantic. As we know, Perkins had a sales agent in the city of Bath where John Haygarth tested the tractor against a wooden facsimile, thereby establishing that its effect, if any, arose not from the properties of the thing itself but from the patient's beliefs and expectations. Given the tractor's reputation for wonder-working, the Perkins phenomenon, like the antidepressant phenomenon, was undoubtedly fueled by anecdote in addition to more formal publicity, and the publicity itself must have set off secondary explosions of enthusiasm. While Josiah Wedgwood perfected such techniques of modern marketing as the celebrity endorsement, the younger Perkins understood the uses of celebrity well enough to cite none other than George Washington as a devotee of the tractor. Perhaps both Perkinses deserve to be remembered among the innovators of marketing for their skill in magnifying a useless object into an international medical fashion.

Wedgwood's use of the great to showcase his wares rested on an explicit theory: "*Fashion* is infinitely superior to *merit* in many respects, and it is plain from a thousand instances that if you have a favourite child you wish the public to fondle & take notice of, you have only to make choice of proper sponcers [sic]."[23] Similarly but in a good republican manner, Perkins recommended his treatment to the world by citing the testimonials of medical men as well as civic figures like the Attorney General of Connecticut. In the case of antidepressants, influential psychiatrists known to the pharmaceutical industry as "key opinion leaders" act as unofficial surrogates of the industry, recommending the drugs to their brethren who then prescribe them widely even if their actual merit, in Wedgwood's terms, is in dispute.[24] In each case, however, and at every level, to follow an opinion leader is indistinguishable from going along with others doing the same thing; the market itself acts as a social mechanism.

While the Perkins tractor was represented as a sort of cure-all and antidepressants are not, the very scale and generality of the market in both cases imply that the treatments are addressed to a range of conditions—extending in the first case from ordinary aches and pains to epilepsy, in the second from ordinary sadness to despondency.

2. Observed Effects, Unknown Causes

While selective serotonin reuptake inhibitors (SSRIs) are used to treat an entire spectrum of conditions from eating disorders to drug abuse, they are most closely associated with the treatment of depression. The medications have been marketed with spectacular success even though the mechanism of depression is unknown—the theory that it is somehow driven by a deficiency of the neurotransmitter serotonin remaining to this day unproven. If the drugs work, the exact reason they work seems moot. The same was said about the Perkins tractor, which, as noted, was also used for a panorama of ailments. While the instrument was shown to work in empirical trials (meaning that some of those treated reported improvement in their symptoms), Benjamin Perkins in his role as the heir, defender, and publicist of the Perkins tractor freely admitted that neither he nor anyone else knew how or why it worked. Indeed, he advertised the tractor's unknown operation as a point in its favor, as showing that verified facts are more important than theories.

> We frequently hear men, whose wisdom is perhaps confined to their significant looks and manner of expression, observe, on a relation of any newly discovered phenomena, that "These things *cannot* be: I know of no principle, or possible operation in nature, by which such effects can be produced." As if the great Creator of the Universe had made no laws relating to the œconomy of nature, which had not been communicated to them, and familiarized to their understandings. I shall take the liberty to observe, as a well known fact, that mere hypothetical reasoning, unaccompanied by experiment, never accurately investigated the properties of any medicine, or predetermined its effects upon the human body.[25]

The author concludes, "We certainly ought not to reject a practice which produces salutary effects, because we do not know *how* it produces them,"[26] a statement that retains its topicality. We have already encountered a contemporary equivalent: "I'm going to have you get some B-12 injections. They have helped many other patients, but I cannot explain to you why they work . . . I can simply say that many patients tell me they feel better and stronger after such a course of therapy."[27] Note that the doctor in this case plays directly on the expectation that the benefits others enjoy, the patient stands to experience as well.

The better to distinguish his tractor from the hocus-pocus of Mesmerism and animal magnetism, the younger Perkins suggested that it had something to do with animal electricity. It is now said that depression results from chemical imbalances in the brain. The fact is, however, that depression can be alleviated by "drugs that increase serotonin, drugs that decrease it, and drugs that do not affect it at all," as well as by sedatives, opiates, and St. John's wort, which argues against the theory that depression is caused by a specific imbalance that SSRI's correct.[28] The phrase "chemical imbalance" has evocative power, as talk of electricity did in the day of Franklin and Galvani.[29]

3. Anecdotal Evidence

The claim that a treatment works because experience shows it works has an irrefutable sound. "Psychiatrists who use [that is, prescribe] antidepressants—and that's most of them—and patients who take them might insist that they know from clinical experience that the drugs work. But anecdotes are known to be a treacherous way to evaluate medical treatments."[30] Perkins's defense of his father's invention consists in essence of a series of anecdotes. He reports, for example, that Mr. Meigs, Professor of Natural Philosophy in the University of New Haven, reports that he successfully treated his son with a Perkins tractor. One minute the boy was in pain and motionless with a high fever; "in about half an hour he declared his pain was gone, turned himself without difficulty on his right side, and fell into a profound sleep" from which he awoke in perfect health. Similarly, Mr. Woodward, Professor of Natural Philosophy in the University of Dartmouth, is reported to report, "I have made use of your Tractors in various disorders, and besides universally abating, and generally removing pains in the head, face, teeth, etc. I have found them useful in the Salt Rheum."[31] One Perkins anecdote tells of the cure of a horse.[32]

It was to correct for unreliable evidence, of which these are extreme examples, that methodologically demanding clinical trials involving placebo controls as well as randomization and double-blinding were put into place some decades ago. But if the marketing of psychiatric wonder drugs recalls an earlier medical fashion, so too does the practice of rigorous verification recall Haygarth's ingenious and telling, if methodologically crude, exposé of the Perkins tractor. Educated in a skeptical tradition, Haygarth was well aware of the fallibility of our experience.

4. Great Expectations

As it happens, clinical trials have revealed much about the mechanism of antidepressants' efficacy. Their efficacy derives in great part not from their specific composition but their way of activating the expectations of those who take them—expectations fed by private hopes, professional recommendations, the official and unofficial mythology surrounding the drugs, and the impression that untold others have already enjoyed their benefits. In one study of antidepressant medication, fully 90% of patients who reported an expectation of improvement duly responded to treatment.[33] That expectation fuels the action of antidepressants is consistent with the findings of placebo research. "Placebo effects illustrate a basic principle of psychological functioning: the self-confirming nature of response expectancy."[34]

This conclusion too was anticipated by Haygarth. Recall his advice to "any person [who] would repeat these experiments" that "it should be done with due solemnity. During the process, the wonderful cures which this remedy is said to have performed ought to be particularly related."[35] Evidently the Perkins tractor was the object of a kind of folkloric cult, to which Haygarth introduced the study subjects in case they were not familiar with it, or possibly to whet their hopes. (On the other side of the Atlantic the instrument was enveloped by such a mist of second-hand information that a number of medical men attested that their experience confirmed the reports they heard of it. It did not occur to them that they might be witnessing "the self-confirming nature of response expectancy.") It was of stuff like circulating anecdotes and shared fascination that the expectation of cure was fashioned, and as Haygarth implies, unless they expected relief the subjects would not have experienced it.

The social and indeed national buzz surrounding antidepressants builds up expectations that play into our very experience of them.

> People now learn about medications through stories and advertisements in print and video media as well as in everyday conversation and they expect the drugs will help them. They have often heard from friends, relatives, the media, and professionals what sorts of psychic and physical changes they will experience from, for example, taking Prozac. These preconceptions affect their actual experience of the drug.[36]

The evidence suggests that the Perkins phenomenon was socially driven in the same way, mutatis mutandis. The tractor created a stir and aroused wonder; people heard about it from neighbors and read about it in newspapers and pamphlets; professionals, at least some of them, recommended it. The social electricity running through the phenomenon prepared the user for the elemental electricity the thing supposedly channeled.

In order to expect the same wondrous effects others were said to enjoy, Haygarth's subjects had to believe that the experiences of others were transferable to them. A set of tractors might be in the possession of George Washington, but it was the cure of more ordinary folk that recommended the tractors at ground level, and presumably it was also such marvels that were celebrated in the "exaggerated stories which, for some months past," circulated in Bath.[37] From nearby Bristol a colleague of Haygarth's wrote that when word spread of the cures he performed with the Perkins tractor (actually it was a sham), he was inundated with patients seeking the same. It was as if Haygarth and his fellow investigators identified not only the mechanism of expectation by which the tractor acted on the patient, but the social sources, the shared imaginings, underwriting its power. The strange sensations reported by those treated with sham tractors were the sort of thing that others like them reportedly felt, a translation into their own bodies of the social energy activating this strange medical movement itself.

Two centuries later, in a different atmosphere of circulating anecdotes and shared fascination, the advertising for antidepressants depicts as suggestively as possible people like the consumer enjoying the benefits of the drugs.

5. Professional Enthusiasm

If the presumed or reported experiences of others recommend a drug, it is a doctor who prescribes it.

Antidepressants have won medical enthusiasm not only because they are in principle more effective and less time-consuming than talking cures, and because the brain is more fashionable just now than the mind, but for the related reason that such drugs conform to the medical model enshrined in the third edition American Psychiatric Association's *Diagnostic and Statistical Manual of Mental Disorders*, which appeared in 1980, only a few years before Prozac itself. Indeed, DSM-III can be said to have contributed to the making of drugs like Prozac in that by medicalizing the language of mental health, it "created enormous professional and financial incentives for both researchers and pharmaceutical companies. It gave them specific diagnoses to target their research and development efforts for prospective treatments."[38] In prescribing antidepressants, physicians may feel that they are in tune with history—employing the very categories that drive the discovery of new treatments, as well as doing what most of their colleagues do—and this belief will communicate itself to their patients.

Haygarth well understood, and stated, that the physician's belief in a treatment contributes to the treatment itself. "In the best manner possible a patient ought to be always inspired with confidence in any remedy which is administered,"[39] by which he did not mean that patients should be hoodwinked, but encouraged to put their faith and trust in bona fide treatments.

As some have noted, a drug is actually not either a medication or a placebo but may be both, engaging the booster effects of hope and expectation in addition to its specific efficacy. Ironically, antidepressants themselves seem to be hybrids, but with little of their own to offer above and beyond placebo in most cases.

It is clear, anyway, that Haygarth envisioned even a cure as a social transaction.

6. Hidden Failures

Particularly damaging to the reputation of antidepressants was the disclosure that most placebo-controlled trials of the most popular such drugs for over a decade beginning in 1987 (when Prozac received FDA approval) yielded a negative result. Because these embarrassments were concealed from public view while the successful trials were highly publicized, the impression formed that the drugs in question were truly effective—and in relieving not only the symptoms of depression but even the lack of self-esteem that according to conventional wisdom lies at the root of unhappiness itself.[40]

At the time of the original research into the placebo effect, the practice of not figuring failures into estimates of efficacy seems to have been common—common enough, anyway, for Haygarth and his associates to distance themselves from it. In his report on the efficacy of cold-water treatments for fever, Haygarth's junior colleague James Currie thought it both more economical and more candid to record only his failures. Falconer, in the preface to his study of the medicinal effects of Bath waters, reproached the practice of concealing failures. Of both Falconer and Haygarth himself it has been said that they "did not yield to any temptation to 'cook the books,' despite their clear vested interest in promoting the taking of the waters in the Bath General Infirmary to patients from outside the city."[41] In his exposé of the Perkins cult Haygarth pointed out that Perkins did cook the books—by not publishing his failures, which were many. "The cases which have been published [by Perkins] are selected from many which were unsuccessful, and passed over in silence."[42] Interestingly, Haygarth does not deny that there were successful cases, only that the success was owing to the composition of the tractor.

It may have been to pre-empt such criticisms that the younger Perkins included in his pamphlet one or two admissions that the tractor treatment wasn't always successful.[43] Such cases are mere specks, however, and the passages in question serve rhetorically to confirm his image as a man of candor with nothing to fear from the facts.

⊟

Surveying the record of psychiatric drugs, a study of placebos in medical history concludes, "Therapies initially are deemed veritable panaceas by patients and enthusiastic healers who describe impressive results. With time, the results falter, skeptical healers report flagging therapeutic efficacy, and new therapies take the place of the older ones."[44] Hence the adage, "Use the new drugs quickly, while they still have the power to heal."[45] With antidepressants revealed as only marginally superior to placebos in most cases and the press now questioning the efficacy of these drugs formerly marketed to much fanfare ("expensive Tic Tacs," *Newsweek* branded them),[46] we seem to be somewhere in the middle of this cycle.

Prozac was once acclaimed as superior to cognitive-behavioral therapy;[47] now some, the defender of the Pollyanna principle among them, rate cognitive-behavioral therapy as superior to Prozac. Many drugs, it seems, have a sort of half-life, a function of their popularity not only with consumers but the doctors who recommend and prescribe them. It is a commonplace of the placebo literature that medications prescribed more enthusiastically work more effectively. It is less commonly noted that medical enthusiasm waxes and wanes with the vicissitudes of fashion, so that "if a new and better drug comes out, the drug it replaces begins to perform consistently less well in tests, merely because doctors have lost confidence in it."[48]

Doctors lose enthusiasm collectively, it seems; they too feel as their fellows feel.

As for the Perkins mania, within a few years of Haygarth's exposé it burned out.

Chapter Eleven

The Power of Rhetoric
Two Healing Movements

The rhetoric driving both Mesmerism and a comparable therapy of our own
time has something to do with such efficacy as they possess.

With its way of activating crusading energies, generating legitimacy and,
especially, expanding the category of depression itself, the campaign
on behalf of depression has been likened to a movement. Here I consider
two healing movements, one of John Haygarth's time and one of ours, both
of which gathered strength as more and more within their orbit came to feel
what it appeared others did. In both cases the originators of the therapy not
only make their case to the world but envision and portray the therapy as
a boon to humanity. In both cases bold claims mixed with reports of success—a compound well calculated to impress—attract audiences and feed
the expectation and the experience of therapeutic effect. In both cases the
most potent advertisements for the therapy may be the testimonials of those
who come forward as witnesses to its power. In both cases, as the therapy
acquires converts and defenders, such witnesses multiply and it grows into
a movement and taps the power of a movement. In both cases, regardless of
its disputed scientific status, the therapy appears to have some effect in some
instances, but, arguably, only insofar as the rhetoric driving the movement
is itself effective. I look into the kinship between the two modes of healing
not to settle the question of their status but to bring out the contribution of
rhetoric to both—especially the more recent—and throw light on possibly
social sources of therapeutic efficacy.

⠿

Not long after post-traumatic stress disorder was recognized in the third
edition of the *Diagnostic and Statistical Manual* published in the wake of
the Vietnam War, the case began to be made that any civilian is potentially
subject to traumas psychologically comparable to the horrors of combat.

With the ensuing expansion of eligibility for PTSD went new therapies, of which one has enjoyed exceptional and perhaps unprecedented popularity: Eye Movement Desensitization and Reprocessing, or EMDR. As unlikely as it may sound, EMDR professes—or originally professed—to treat PTSD by rapid shifting of the eyes under the guidance of a trained therapist. EMDR has been acclaimed by the media (from TV to *Stars and Stripes*), endorsed by figures affiliated with the American Red Cross, the FBI, UNICEF, and the Menninger Clinic among other bodies, and recommended in clinical guidelines published by the American Psychiatric Association and the Veterans Administration. A book on EMDR may in fact preface the text with pages of testimonials, reminding the reader that EMDR is not just a method but a movement and inviting him or her to experience the moral electricity that animates it, to feel what others feel. With some 70,000 practitioners around the world,[1] EMDR claims cures in the millions, although the mechanism by which it works, if work it does, remains unknown. The large claims originally made on behalf of such an obscure technique as eye-shifting have prompted comparisons between EMDR and Mesmerism, which as we know was brilliantly promoted in the twilight years of pre-revolutionary France as a way to cure any and all ills by channeling the mysterious force known as animal magnetism.

In an article published in 1999 Richard McNally detailed seventeen parallels between EMDR and Mesmerism, most concerning the way the two movements were launched, promoted, and defended against critics. Thus, for example, "Both Mesmer and [Francine] Shapiro [the founder of EMDR] had nontraditional backgrounds and entered the mainstream of the field from its periphery"; "both animal magnetism therapy and EMDR have been applied to an astonishingly wide range of conditions"; "both Mesmer and Shapiro have claimed that 'Establishment' clinicians have been biased against their therapies."[2] Accordingly, EMDR, like Mesmerism, is assigned to the dubious category of therapies that spring up on the fringes of the field and make assertions about their own revolutionary potential, in this case the potential to eliminate human suffering, that measure their distance from the scientific center. McNally does not deny that these treatments may work, only that they work by some means other than the power of suggestion. My claim, compatible with his, is that the rhetoric driving EMDR and Mesmerism, along the dynamism of the movements themselves, enhances their suggestive power. Both therapies speak to their time and place and otherwise use rhetoric to their advantage. Proponents of both cite empirical observations and attested experience, which puts those who dismiss the therapies as fanciful in the awkward rhetorical position of having facts against them. And much as Mesmerism itself was magnetic in the sense that it attracted both fascination and followers, the EMDR movement itself acts like a dramatization of the "natural movement toward health"[3] that EMDR professes to enable. Both systems realize an abstract, even recondite doctrine in vivid social ways, thereby enhancing their persuasiveness.

Promoted by its adherents as the remedy not just for PTSD but a long list of ailments, and embraced by an international following, EMDR has leanings to the universal. Mesmerism was in touch with the very fluid of the universe: animal magnetism. Though the channeling of an essence as subtle as animal magnetism is not something that really lends itself to explication, the theory of Mesmerism held that disorders arose when the fluid somehow became obstructed or unbalanced in the body.[4] Mesmer, a physician by training (his dissertation concerned planetary influences on disease), professed to cure ills by discharging the blocked power, in the process throwing people into expressive convulsions that did not fail to attract commentary and ridicule. It was to free the flow of magnetic fluid that mesmerized subjects were arrayed in "chains"; hence satiric depictions of a mesmerist session as an assemblage of bodies, each somehow touching another. Evidently the notion that they were blocked and the blockage could be overcome by a current of fluid made sense to Mesmer's subjects. Mesmerism had the attraction of a novel doctrine, one very much in tune with the 1780s, overlaid on the intuitive model of an imbalance corrected or an excess cleared. It could claim at least two advantages over bleeding and purging: first, that it addressed itself to a force more profound and fundamental than gross fluids, and more modern than the traditional and always rather academic humors; second, that it could direct this force without subjecting the patient to physical discharges. And no sooner were Mesmeric cures performed than they were written up in pamphlet form, to join the other writings that surrounded the movement like a buzz. It is said that the published accounts of such cures circulating in France "must have sapped the faith of many Frenchmen in the purgative potions and bleeding used by conventional doctors."[5]

In the decade before the French Revolution talk of Mesmerism seemed to fill the air much as animal magnetism itself—the object of general fascination—was said to fill or underlie all things. As noted, the ability of Mesmerism to be everywhere and to stir everyone must have read like a dramatization of the fundamental power itself. If one could not observe the magnetic fluid directly, nevertheless one could feel the attraction of the movement that was in touch with it, a movement whose magnetism acted like a confirmation of its own theory. Moreover, the prospect of liberating a blocked elemental fluid appealed to the Enlightenment understanding of Nature itself as an order prior and superior to all artifices and obstructions. A classic study of Rousseau is subtitled "Transparency and Obstruction" in token of the subject's belief that in a natural state one heart reveals itself to another, and yet we find ourselves in a world where sight is blocked and hearts closed.[6] It is because he believes that things have gotten twisted out of their original shape in this way that Rousseau's rhetoric is full of stunning paradoxes. To see society through Rousseau's eyes—and he himself both invented and perfected the role of the unconventional genius who challenges the truths of the center, in this case Paris—is to see it as something like perverted potential or blocked energy, as many no doubt do even now.

A couple of generations ago an eminent literary critic commented on "the half-baked Rousseauism in which most of us have been brought up,"[7] and the enduring magnetism of Rousseau's rhetoric may give us some sense of that of the mesmerist movement and the pamphlets that were its chosen medium of expression (some two hundred appearing in the 1780s).[8] If mesmerist writing was marked by a "tone of injured innocence and opposition to the . . . establishment,"[9] so was Rousseau's, after all, except that his quarrel extended beyond the citadels of science; and if Mesmer came up with Mesmerism after sojourning for some time in a forest, Rousseau himself "wandered deep into the forest" in composing the work where he "dared to strip man's nature naked,"[10] the *Discourse on Inequality*. Not only did mesmerist cosmology serve as a seemingly apolitical vehicle for Rousseauist ideals in the intense atmosphere of the pre-revolutionary decade, but the dramatic rituals of Mesmerism (made all the more so by the maestro's robes and wands) seemed to demonstrate that only by cutting through the customs of polite society could health be restored. Little wonder that an experience so charged with ambiguous suggestion and so potentially subversive incurred the suspicion of authorities, including the King of France himself, who in 1784 appointed a commission of eminent scientists including the skeptical Benjamin Franklin to look into the phenomenon.

Having induced the effects of Mesmerism in subjects who were not magnetized but believed they were, and having found that subjects exposed to the alleged magnetism without knowing it remained unaffected, the investigators concluded that animal magnetism had no existence. The therapeutic power of Mesmerism was thus exposed as an artifact of what is now known as the placebo effect, though we might also consider it a social artifact. A thoughtful commentator on the placebo effect emphasizes its social character thus:

> Increasingly it is hard to deny that giving placebo has a very important therapeutic effect or that being studied, *participating in a group*, is highly beneficial. The implications of this effect for joining groups are obvious. Humans are social animals. . . . . Talking gives permission to act, sometimes—[11]

as the rhetoric and ritual of Mesmerism licensed behavior that would be unimaginable otherwise. It follows that the magnetism of the mesmerist movement, its powerful appeal to our social nature, contributed to its therapeutic effect.

In his blindness to the merits of EMDR McNally has been likened to Benjamin Franklin, a paradoxically complimentary reproach that suggests a connection between the two therapies.[12] According to Mesmer, sickness is caused by an obstructed flow of magnetism. According to the founder of EMDR, psychological ills are caused by blockages in the nervous system. "The system becomes 'stuck.'" Traumas "remain locked in the person's nervous system"; or less positively, "the inner state experienced during the traumatic event is apparently locked in the victim's nervous system" (pp. xiv, 66, 182). Healing takes place when the system is unlocked and the obstruction cleared away, a process that completes itself in short order, all but automatically (provided the proper steps are followed), once EMDR is initiated. The "wonderful change in his nervous system" denied Mr. Woodhouse at the end of *Emma* thus comes to the EMDR client. Where Mesmer's vats and iron rods acted as a visual rhetoric demonstrating the physical nature of an ethereal fluid, EMDR employs stimuli like hand-taps and darting lights consistent with the allegedly neurological basis of the fateful blockages; where Mesmer spoke of "poles, streams, discharges, conductors, isolators, and accumulators," EMDR theory posits the storage of negative memories "in a neuro network with a high bioelectric valence associated with the high level of dysfunctional affect."[13] Like Mesmerism, which portrayed itself not as a romantic alternative to science but science itself, EMDR grounds itself in the "laws of cause and effect" (p. 242)—and has spawned much neurobiological jargon—but remains a mystery. Not in question is EMDR's character as a movement; over the 1990s it not only established itself in the United States but was taught in training sessions from Japan to South Africa, from Australia to Brazil. If participating in a group can be of therapeutic value in and of itself, all the more is this true of participation in a crusade, and one dedicated to explicitly therapeutic ends at that. The utopianism of the healing movement from which psychotherapy evolved[14] returns in EMDR.

Where Mesmerism fascinated a France that was also fascinated with electricity and Franklin immortalized himself as the man who captured lightning, EMDR for its part is lightning in a bottle: a therapeutic method equally swift and powerful, so it is said. Like Mesmerism, EMDR claims to cut right to the heart of things; it is direct and dramatic, indeed spectacular in its own way, as befits the release of pent-up energy. Its stories tell of victims seemingly locked in suffering until an exposure to EMDR summarily cures them and returns them to life, quite as if a jammed mechanism had been freed up or a reflex triggered or, indeed, an obstruction removed. And by analogy with the notarized pamphlets documenting Mesmeric cures, EMDR literature certifies these stories by citing endorsements and praises. In both cases individual cures are swept up in a larger narrative of healing and transformation. As Mesmer, upon his return to civilization from the forest, vowed to "pass on to humanity . . . the inestimable benefaction that I had in hand,"[15] so has the founder of EMDR offered humanity a powerful boon "that might lead to the eventual healing of us all" (p. 242)—although both donors tried to keep control of their bequest, in the one case by not divulging the mystery or divulging it only to subscribers or declaring it sacrosanct and unalterable once it had been divulged; in the other case by licensing initiates. As with Mesmerism, too, EMDR has inspired rhetorical avowals of its world-changing potential. "Claims of global historic significance have been made on behalf of both Mesmerism and EMDR."[16]

Many EMDR cures appear to turn on the retrieval of a buried memory, and indeed EMDR grew up in tandem with the repressed memory movement—another instance of social synergy. The nervous-system blockage both postulated and allegedly overcome by EMDR is something very like such a memory, now given a scientific-sounding description and made to seem entirely uncontroversial. Like Mesmerism restoring the harmonious flow of animal magnetism, "EMDR can remove the block that is preventing the natural movement toward health. It can release you into the present you always wanted for yourself, a present where you can feel free and in control" (pp. 11-12), and it is prepared to offer one anecdote of recovery after another to substantiate this promise (as Mesmer carried around written testimonials to his power). Like the citation of endorsements, the multiplication of stories conveys the impression that EMDR is not an abstract doctrine but a living movement—one the reader is invited to join. "Humans are social animals." Repetition, perhaps the principal figure of rhetoric, serves both to emphasize just how much the EMDR movement is capable of and to make the mysterious neurological process of unblocking seem not familiar, even intuitive.

Committed to the theory that PTSD and similar conditions actually result from some kind of blockage, EMDR rhetoric employs the metaphor of obstruction with an insistent literalism. Repetition establishes what might otherwise seem strained, as in this characteristic passage addressed to the reader in the original EMDR manifesto:

> When you cut your hand, your body works to close and heal the wound. If something blocks the healing . . . the wound will fester and cause pain. If the block is removed, healing will resume. A similar sequence of events seems to occur with mental processes. That is, the natural tendency of the brain's information-processing system is to move toward a state of mental health. However, if the system is blocked or becomes imbalanced by the impact of a trauma, maladaptive responses are observed. . . . If the block is removed, processing resumes and takes the information toward a state of adaptive resolution and functional integration.[17]

It turns out that any number of questionable presumptions are packed into this seemingly straightforward, but actually completely speculative, model of information processing, some version of which remains to this day the official foundation of EMDR. It is implied, and EMDR contends, that the mind heals like the body and should heal at least as quickly, that the mind's healing response can be stimulated by physical means, that any therapy that does not address itself to the theorized underlying cause of the maladaptive responses will fall short, that conversely EMDR succeeds because the pathway to self-forgiveness, self-affirmation, and similarly desirable states is wired into us and EMDR activates it, and that the memory of the original trauma, being locked into the nervous system, does not alter over time. ("This information is stored in the same form in which it was initially experienced, because the information-processing system has, for some reason, been blocked."[18]) Though questionable from top to bottom, the model of a blockage overcome has a seeming transparency well suited to its function of making EMDR believable. One does not join a movement without believing in it. Indeed, it was in 1991, when EMD was renamed EMDR in accordance with the founder's belief that it was really an information-processing therapy, that it began to take on the identity of a movement, spawning institutions and spreading to other continents.[19] Only when eye-movement therapy found the right rhetoric did it take wing. And just as crusades feed on themselves, so Eye Movement Desensitization and Reprocessing continued to thrive even after disclaiming the necessity of eye movements.

It is with the aid of physical stimuli that EMDR accomplishes the freeing of blocked energy. Like the theory of blockage itself, which solicits belief because it is so straightforward, the use of physical stimuli appeals to our love of directness. EMDR speaks in a sort of populist idiom that favors the literal and the immediate over the ambiguous or the indirect, and its physical exercises are the props of its rhetoric. The signature EMDR technique of shifting the eyes as if wiggling or loosening something stuck seems like a literal application of the theory that to overcome the memory of trauma we need only free up a mechanism. (Similarly, the technique of tapping, or "tactile stimulation," resembles what we might do to produce a knee-jerk.) Mesmer, though he liked to attach patients to each other to form circuits, avoided "knots, which created obstacles" to the flow of magnetism.[20] EMDR appears to take its own theory of obstruction no less physically.

Whatever else may have been going on in mesmerist sessions in pre-revolutionary France, they were occasions of license, which is one reason the authorities viewed them with suspicion. Roping themselves together, going into fits, breaking out in laughter, Mesmer's willing subjects seized the possibilities of license and behaved in ways they ordinarily would not. While EMDR training sessions have reportedly witnessed some strange behavior, EMDR per se seems to offer a message of absolution embedded in a series of actions that function like a rite. If Mesmer acted as the ministrant of a universal power, the EMDR therapist performs a minutely specified set of procedures—a sort of priestly ritual concluding with a "body scan" for any remaining physical traces of trauma—the effect of which, in theory, is to clear away all "negative cognitions." The therapist thus does Mesmerism one better, placing him- or herself *en rapport* with the client morally and authorizing not bizarre behavior on a special occasion but the client's very self. "My subjects' insights had followed their own logical (and emotionally healthy) train of thought, moving, for example, from 'I was to blame' . . . through 'I did the best I could,' and finally to 'it wasn't my fault. I am fine as I am'" (p. 26).[21] That EMDR-seekers are fine as they are is the axiom and the conclusion, the presumption and the end-point, of EMDR. The client comes to the predetermined insight, "I deserve love. I am a good person. I am fine as I am. I am worthy; I am honorable. I am lovable. I am deserving (fine/okay). I deserve good things," etc.[22] In a variation on the principle that talk authorizes action, EMDR offers permission to be ourselves. It is any case a way of speaking as much as a method, its efficacy bound up in the oft-repeated claim that we need only follow the healthy tendency of our nervous system, our physical nature, to realize we are worthy of love. Only when and where this line of argument resonates will EMDR possibly work.

In its stronger form EMDR maintains that precisely because the patient's nervous system is locked, merely verbal therapies are useless; in this sense it portrays itself, in the spirit of Mesmerism, as a uniquely potent method of healing, not just one method among others. In its weaker form EMDR holds that protracted therapies are usually unnecessary—though even on this showing it remains remarkable. One goes with astonishing speed from being locked in a state of trauma to being "entirely free of emotional turmoil" (p. 135), as by turning a key. Quite simply, "a person's internal information-processing system is stimulated so that the core of health that is within can blossom forth" (pp. 135-36; cf. e.g. pp. 25, 29). Just as references to information-processing sound good in an age swept up in an information revolution (the same revolution that has powered the expansion of EMDR itself, now with multiple Web sites and a voluminous literature), so EMDR's theory that psychological suffering results from emotional imprisonment, and healing means walking out of the cell, has considerable rhetorical appeal in this time and place. Rhetoric—speech as an instrument of action—is indeed the mobilizing force of the EMDR movement, as we are reminded when the founder of EMDR repeats evocative phrases, declares that her own book "allows us to . . . celebrate the triumph of the individual" (p. 29), exalts the mission of EMDR, or simply addresses the reader:

> Now there is reason to hope. EMDR is not a panacea, but it may be able to unlock your innate, physiological healing system and allow you to change at a rate and in a way you never thought possible. (p. 12)

Mesmerism caught on in a France intoxicated with the wonder of balloon flight and left to speculate where the line between the possible and the impossible might actually lie. Was it possible to magnetize a tree? To revive a dead dog? EMDR literature fosters a sense of expanded possibility by portraying EMDR as a way of bringing emotionally frozen people back to life.

Is this believable? Can people be cured as automatically and definitively as they are in the cases recounted in story form in *EMDR*—stories that are really reconstructions centering on the reports of persons whose names have been changed? EMDR stories are too uniform in their unfolding and too Cinderella-like in their outcome, as well as too dependent on the dubious practice of retrieving repressed memories, to command full belief. (For Tolstoy, who thought deeply about stories, "a story that is too neat, explains too much, and makes the world all of a piece must be false."[23] EMDR stories satisfy all three criteria.) On the other hand, there are too many testimonials to EMDR from too many sources to doubt that it constitutes a movement and as such has an inspirational effect—all the greater because of its advertised character as a rescue mission. If, as the investigators of Mesmerism concluded in 1784, "Man has the capacity to act on his peers, to shake their nervous system to the point of convulsions, without the help of any fluid,"[24] so can EMDR console and inspire without administering the neurological shake prescribed by its own theories.

Where a movement seems to build on itself, rhetoric—the driving force of a movement—builds on themes and tropes already familiar, and so it is with EMDR. Around the time the founder of EMDR wrote of a victim of trauma, "EMDR had jump-started Linda's own healing process" (p. 4), there appeared a book provocatively titled *The Placebo Response: How You Can Release the Body's Inner Pharmacy for Better Health*.[25] The idea that we possess a healing power that can somehow be activated from without was in motion; the idea itself had power. EMDR adds that the power awaiting release, in this case belonging to both body and mind, may somehow get tied up. When a trauma remains unprocessed, "the system becomes 'stuck'" (p. 18). "The problem many trauma victims face is that the upsetting experience from their past . . . is 'stuck' in their nervous system" (p. 23). As the scare quotes around *stuck* may suggest, the term belongs to our cultural vernacular. Robert Pirsig's *Zen and the Art of Motorcycle Maintenance* contains discourses on getting stuck and unstuck; Billy Pilgrim in Kurt Vonnegut's *Slaughterhouse Five*, another cult classic, comes "unstuck in time." EMDR's founder seems to believe that psychology was itself stuck until EMDR. "We went from Kitty Hawk to a man on the moon in little more than 50 years, yet we have not had a major paradigm shift in psychology since Freud, nearly a century ago."[26] However, since it entered our common lexicon around the time of Pirsig and Vonnegut, the notion of a paradigm-shift has become a received idea in its own right.

In point of fact, the paradigm-shift proclaimed by EMDR is an exaggeration. Just as rhetoric in general plays on commonplaces (resonating precisely because it does so), so the rhetoric of EMDR trades heavily on the formulas of pop psychology. Ever since the 1960s, after all, the self-help movement has been marketing step-by-step exercises for breaking the hold of the past and reprogramming the self, always with the presumption that self-blame is poisonous and we are fine as we are, just as in EMDR.[27] If we want to know what "I am fine as I am" really means, for example, we might consult *Compassion and Self-Hate* (1975; reprinted 1998), featuring the following proclamation:

> The fact of my being is enough. I require no terms, conditions or permits from myself or anyone else. I live, and in living I am fully entitled to go on living. My life, my existence, my being is not predicated on standards, values, achievements, or accomplishments. . . . I must fight to give myself the right to feel good about myself and to feel good mood-wise, regardless of any accomplishment or non-accomplishment whatsoever.[28]

EMDR's novelty is to underpin the rhetorically asserted "right to feel good about myself" with the science, if that is what it is, of blocked energy, thus grounding the right in our physical being and proving that everything celebrated by pop psychology is actually already ours. According to EMDR, that is, the nervous system itself favors such "positive cognitions" as "I am fine as I am. I am worthy; I am honorable"; our very bodies are constructed in accordance with the dictates of EMDR. Therefore, it is implied, EMDR clients who arrive at the insight that they are fine, worthy, and honorable respond not to suggestion but to the prompting of their physical selves. Far from breaking with everything that came before, EMDR taps deeply into received ideas, prominently including the right to be oneself even in a world hostile to the self. Unless these doctrines were already in place, EMDR would not be able to cite them as first principles; they would sound less like self-evident truths than like utopian propositions. (More's Utopians are indeed portrayed as fine, worthy, and honorable, precisely because they belong to the same excellent commonwealth.) If EMDR speaks a sort of pop idiom, affirming the "innate wisdom and health" of every one of us (p. 25), we can now identify the idiom as a variant of that of pop psychology. One movement builds on another.

Like EMDR, the placebo effect has been theorized as mobilizing an innate capacity for healing and bridging body and mind, and celebrated as heralding nothing less than a paradigm-shift.[29] Arguably, in fact, the success of EMDR lies in its mobilization of the placebo effect. In one study PTSD was effectively treated by a placebo therapy featuring an empathic listener, an impressive name ("present-centered therapy," or PCT), and a certain methodology, but "lacking the specific ingredients postulated to be necessary" for the successful treatment of the condition.[30] There is every reason to believe that EMDR, with its humanitarian ethos, technical-sounding name and elaborate, almost priestly methodology, is just such a placebo, enhanced with considerable social power. Rounding out the analogy between Mesmerism and EMDR, then, is that the former inspired the first concerted investigation of the effect tapped by the latter. As we read of one organization after another, including the FBI, that now recognizes EMDR, we get the sense of a mesmeric chain around one of the maestro's vats of magnetized water. A circular chain may also serve as an image of the theory and practice of EMDR: evocative rhetoric drives a movement that contributes to the very efficacy of EMDR (in that group membership empowers, inspires, "is highly beneficial"), which in turn yields stories that feed back into the movement's rhetoric. But in order for rhetoric to be evocative, it must be attuned to its time and place. EMDR would not have caught on in, say, the 1950s because at that time doctrines like "I am fine as I am," along with a host of correlates, had not yet been established. Nor, for that matter, had the diagnosis of PTSD been formulated; too much of a stigma still attached to the idea of a disorder to allow for a disorder caused by things completely beyond one's control.[31] (Thus too, while EMDR has been introduced into Japan, one wonders how it fares in a culture to which its tenets are ill adapted.) Similarly, if the rhetoric of EMDR were to fall from favor, in all probability the efficacy of EMDR would wane. Only a few years after Paris was swept by mesmerist fervor, Mesmerism fell more or less flat in England; it was too tainted by association with the wildness of the French Revolution for English liking. Time and place were not right.[32]

Before the Revolution, by contrast, Mesmerism found a very favorable climate in France. Powerful suppositions had to be in place in order for it to make sense to so many people, and to move them, and so they were. The suppositions were those of "sentimental empiricism," the philosophy centered on the principle "that feelings were responses to a world outside the mind and were therefore the bedrock of natural knowledge." It was this world-view that underwrote Mesmer's claim that the strong feelings convulsing his patients were responses to an actual fluid, the fluid that constituted the very medium of sensibility. Far from being a homespun doctrine, moreover, sentimental empiricism was the dominant philosophy of natural science at the time. By no means was Mesmer alone in theorizing the existence of a universal medium or envisioning Nature as a single linked entity; his ideas possessed considerable resonance and plausibility, even for many members of the French establishment. "Mesmer's theory is not so much a departure from credible philosophy as an exaggeration of it."[33]

While EMDR, for its part, echoes the familiar language of pop psychology, it also presents itself as science, accruing considerable rhetorical power as a result. Indeed, it too has roots in a credible source: the authoritative directory of mental disorders, the *Diagnostic and Statistical Manual*. Before the founder of EMDR made the claim that "three-quarters of the general public will experience an event that could cause a traumatic response sometime in their lifetime" (p. 176)—an example of the sort of diagnostic inflation that has become a commonplace of discourse, with one claim somehow supporting another—the criteria of traumatic exposure had already been broadened in *DSM-IV* (1994) to the point that merely hearing or learning about someone else's trauma constituted a possible traumatic event in its own right.[34] PTSD originally appeared in *DSM-III* (1980), the edition that for the first time employed diagnostic criteria framed on a medical model. But how did PTSD get into *DSM-III*? It was lobbied into it by psychiatrists who opposed the Vietnam War and now stood in sympathy with the veterans who, they said, had been traumatized by the experience of war. These psychiatrists

argued that many veterans continued to suffer severe stress symptoms long after having returned home. . . . Because there was no place in the existing diagnostic system for either a chronic stress syndrome or a delayed one, these psychiatrists lobbied for inclusion of "post-Vietnam syndrome" in the forthcoming third edition of the *Diagnostic and Statistical Manual of Mental Disorders*. . . . Members of the *DSM-III* task force were reluctant to endorse a diagnosis tied specifically to a historical event. Yet they eventually relented when veterans' advocates persuaded them that the same stress syndrome occurred in survivors of other traumatic events, such as rape, natural disaster, or confinement in a concentration camp. Converging clinical evidence, pointing to a common syndromic consequence of trauma, clinched the inclusion of PTSD in *DSM-III*.[35]

Once installed in *DSM-III*, PTSD came to life, and not only in the pages of the professional literature. The disorder acquired a lay as well as a professional constituency, perhaps because it established for the first time that the consequences of traumatic exposure befell people through no fault or failing of their own. "This is an uncommon situation in psychiatry. [With one or two exceptions] there is probably no other psychiatric diagnosis that has so closely met lay people's and professionals' expectations."[36] By the time EMDR emerged as a treatment of PTSD, the innocence of the victim was an established principle and the disorder itself the subject of a burgeoning literature as well as an expanding definition. The moment was right.

The crafter of the original definition of PTSD in *DSM-III* confirms that "an active group of advocates were lobbying for the inclusion of a diagnosis" that would take account of the trauma of Vietnam veterans. Once instated in *DSM-III*, she writes, "the concept of PTSD took off like a rocket"[37]—the twentieth-century equivalent of the wonder of balloon flight. The implied analogy of advocacy to rocket fuel may give us some notion of the power of rhetoric available to the founder and followers of EMDR. They did not fail to use it. In its overtly rhetorical appeals on behalf of and at times to the traumatized, EMDR recalls the advocacy that constituted PTSD as a medical entity in the first place and later enlarged its boundaries. Indeed, the power of rhetoric that made a cause célèbre of a psychiatric diagnosis is the power applied by EMDR for therapeutic ends.[38]

Chapter Twelve

From Medicine to Psychotherapy
The Placebo Effect

Psychotherapy is a preserve of the placebo effect.

While it is said that medical history until recently is a chronicle of the placebo effect,[1] that doesn't mean the use of placebos died out with the medical innovations of the twentieth century. On the contrary, placebos in the form of distilled water, bromides, vitamins, and the now-infamous sugar pill were administered by doctors at their own discretion well into the century. As late as 1964, it was estimated that somewhere between 20% and 40% of prescriptions were for placebos.[2] In order to evaluate the efficacy of new drugs and treatments, the practice of discounting for the placebo effect has since been built into clinical trials that have become the norm of medical research. Formerly a ruse to be practiced at will, the placebo became a control in a study. Yet if the introduction of the double-blind trial to monitor the placebo effect and establish a drug's efficacy "above and beyond placebo" marked a new phase in their use, to many placebos have come to represent more than dummy treatments that activate a capacity for delusion.

Over recent decades everything about the placebo effect including the sugar-pill model itself came under challenge. Is the placebo effect nothing but a sham? How does it happen that officially inert medications can produce not only felt benefits but even physiological changes? Is it not closer to the truth to say that the body possesses resources for healing that the rituals of medicine tap? Questions like these animate recent literature on the placebo effect, which has become an object of research interest in its own right as well as a topic of general fascination. But for all the reaction against the reduction of the placebo effect to the dimensions of a sugar pill, its reputation, at least in medicine, has not been fully rehabilitated. It remains true that "even when ... physicians are convinced that impressive forces may be rallied through [the placebo effect], they often cannot shake themselves free of the conviction that this practice is at best unreal and at worst chicanery."[3] "Most doctors dislike the idea of the placebo and like to discuss it even less," wrote two Belgian doctors in 2012.[4]

Although it is common knowledge that placebos are still used in medical practice (generally, however, in the form of ordinary analgesics, vitamins, unnecessary antibiotics, sub-therapeutic doses of medication, and latterly "probiotics," not the likes of sugar-pills),[5] they cannot be prescribed as freely as they once were. Decades into the era of informed consent, doctors are simply not at liberty to act as if their former prerogatives had never been called into question. Paradoxically but tellingly, it seems that many of those who prescribe placebos in one guise or another believe it is unethical to do so in clinical practice.[6] Some pharmacies are unwilling to fashion placebos.[7] But if we understand the placebo effect as a benefit arising from a treatment not specifically effective for the condition in question, then not all applications of the placebo effect necessarily involve deception. This chapter argues that even as the routine use of placebos in clinical practice lost its traditional status as an exercise of medical discretion, the placebo effect in the form of suggestion flourished in the practice of psychotherapy; that the robust exercise of the placebo effect, at a time when medicine was becoming more impersonal and more uneasy with the effect itself, enhances the experience of psychotherapy; and that even though the therapist engaged in a talking cure is not to be confused with a medical doctor knowingly administering a sham treatment, the epistemological foundation of psychotherapy is questionable. The emigration route of the placebo effect is sketched out every time the argument is made that because the efficacy of antidepressants is so suspect according to the canons of evidence-based medicine, the depressed are better served with psychotherapy—even though the latter itself may simply be "the quintessential placebo."[8]

Between 1975 and 1990 the number of clinical psychologists in the United States almost tripled, while the population of psychotherapists of other sorts increased even more.[9] Arguably, the explosive growth of psychotherapy over the last generation or two has much to do with the uniquely rich habitat for the placebo effect provided by the institution, and this just when the use of placebos in medicine fell from grace. The placebo effect is exploited more freely—with less reservation and constraint—in psychotherapy than in medicine. Especially given the potential instability of intimate social bonds in postmodern society,[10] the bond with a therapist—the therapeutic alliance, as it is called—is reassuring in and of itself, regardless of the content of therapy. In addition to its successful command of the placebo effect, however, patients entering this now popular institution could take encouragement indirectly from one another, as members of a virtual movement. Perhaps this multiplier phenomenon, whereby the power of the placebo draws multitudes who then exert a social effect of their own, helps account for the charisma investing this postmodern mode of healing.

⊜

Assuming the placebo effect is a benefit (1) derived mainly from the expectation of benefit and (2) registered in the form of feeling better, then psychotherapy that centers professional attention on the patient in the interest of helping him or her feel better is very likely to engage it. As the most comprehensive and searching study of its kind puts it, "Psychiatry and psychotherapy are rife with placebo effects."[11] But where such effects can be distinguished both in theory and practice from the clinical effects of drugs—hence the methodologically demanding trials pitting drug against placebo—they are so woven into the practice of psychotherapy as to complicate the attempt to differentiate them from less impressionistic benefits even in principle. "The main problem in studying placebo effects in psychotherapy is that it is difficult, maybe impossible, to separate the placebo component from the specific effect of a psychotherapy."[12]

According to a notable article that appeared in *Psychological Bulletin* concurrently with the mid-20th-century surge of interest in the placebo effect, "Certain general aspects of the psychotherapeutic relationship seem very similar to those responsible for the so-called placebo effect, which is well known to investigators of the therapeutic efficacy of medications."[13] One of the authors of this seemingly compromising admission went on to publish the landmark *Persuasion and Healing*, where the point is confirmed, for good or ill, by case-histories of patients led to insights about themselves that are believable and encouraging but possibly false. "To be effective, interpretations, the primary means of transmitting the therapist's conceptual framework, need not be correct, only plausible."[14] Unlike a medical doctor carrying out a sham procedure, the psychotherapist on this showing need not disbelieve in proffered interpretations that satisfy the patient but may be quite untrue. And if the healer who is not just an actor but believes in his or her words and deeds makes an especially effective conduit for the placebo effect, then the therapist committed to a "plausible" interpretation is just that.

Even if the person in therapy improves, the improvement is not necessarily a consequence of the therapy. Given that people tend to enter therapy when they hit bottom, "their psychological states at the time . . . are so poor that it is far more likely their mental health will improve than that it will decline, even in the absence of therapy."[15] Many of our troubles pass of their own accord, or when the crises that give rise to them pass. In short, to assess the effectiveness of therapy we would need to take account of factors like regression to the mean and spontaneous remission responsible for the inflation of the placebo's power ever since Beecher omitted to factor them into his estimate of it. But even without these accretions, the placebo effect has plenty to work with in the setting, form, atmosphere, and content of psychotherapy. Just as some portray the administration of placebos as a mode of psychotherapy,[16] so—to complete the union—does psychotherapy itself employ and exploit the placebo effect. The principal author of *Persuasion and Healing* went so far as to portray psychotherapy as a sort of placebo institution, contending that "With many patients the placebo may be as effective as psychotherapy because the placebo condition contains the necessary, and possibly the sufficient, ingredient for much of the beneficial effect of all forms of psychotherapy. This is a helping person who listens to the patient's complaints and offers a procedure to relieve them, thereby inspiring the patient's hopes and combating demoralization."[17] It is presumably because of this inspirational effect that diverse modes of psychotherapy seem to work equally well even though founded on different postulates. Just as medications with different, even contrary, modes of action work against depression because they all tap the placebo effect, so do different modes of psychotherapy conscript the same effect. "The positive effects of therapy have relatively little to do with the specific interventions of the therapist and come largely from nonspecific factors."[18]

Not only is the psychotherapeutic relationship itself patently loaded with placebo potential, but its nature rules out the double-blinding built into clinical trials such as one that recently found vertebroplasty no more effective than a placebo. "Psychotherapy studies cannot be made blind in the manner of placebo controlled medical studies. Quite obviously the therapist must be aware of the treatment being delivered to follow the treatment protocol."[19] Questioning the applicability of the randomized clinical trial—the gold-standard of verification—to psychology, a former president of the American Psychological Association has dismissed randomization and rigorous controls, as well as double-blinding, as "niceties"[20] and contended that it simply doesn't matter that common modes of psychological treatment have not been validated experimentally. It is hard to imagine a medical doctor showing quite this insouciance toward evidence, whatever his or her degree of enthusiasm over evidence-based medicine. One reason psychotherapy is "rife with placebo effects" is that no effort to account for them, comparable to the effort to distinguish the placebo component of medical treatments, has been or perhaps could be made; or to put it the other way around, psychotherapy is so rich with placebo effects that it would be exceedingly difficult to isolate and test critical variables independent of them. Both in psychodynamic and cognitive-behavioral therapy, "patients apparently respond to something more general than any particular theory implies. . . . The quality of the therapeutic alliance largely accounts for the effects of any therapy."[21] Given its dependence on placebo effects, psychotherapy can hardly afford to subject them to the kind of suspicion in which they are still commonly held in medicine. Some argue, accordingly, that the dubious reputation of the placebo in medicine should not be allowed to cast a shadow over psychotherapy.[22]

Regarding the placebo not only as a confounder in clinical trials but a powerful x with a dubious past and an uncertain place in clinical practice at this hour, and a riddle insofar as it mimics physiological responses, medicine today is disturbed by it in a way psychotherapy is not. Psychotherapy does not have medicine's commitment to the model of specific causes and mechanisms and does not have to grapple with such a disconcerting enigma as effective sham surgery (the placebo treatment in the vertebroplasty trial among others). Unlike those physicians who once pretended to treat the patient's body while actually attempting to treat the mind, the psychotherapist can treat the mind in all frankness. Neither, therefore, does psychotherapy have medicine's troubling memory of its own use of the ploys we call placebos— ploys that seem innocent one moment but indefensible the next; producing responses now imaginary, now bewilderingly potent. "The entire enterprise of medicine must necessarily find the notion of placebo effects at the least uncomfortable."[23] Interestingly, the authors do not say the same of psychology even though they write in *The Journal of Clinical Psychology*.

When a specific mode of psychotherapy is tested head to head against a generic therapy in the manner of a drug tested against a placebo, the generic therapy lacks the ingredient in question but includes empathy, attention, support, and other "common factors." In other words, what some call the placebo treatment features the cardinal virtues of the profession itself. Indeed, "psychotherapy might be nothing more than good human interaction between patient and therapist, so that trust, belief, expectation, motivation, and hope, that are common in all types of psychotherapy, would be the factors responsible for the successful therapeutic outcomes."[24] There is thus good reason why psychologists should be well disposed toward the placebo effect even if they don't like the term; and being so disposed, they have come to its aid now that it has fallen from favor in medicine.

⊟

If the placebo effect encompasses a spectrum of responses ranging from the benefits of sham procedures like "Tractoration" all the way to physiological changes resulting from officially inert agents, little wonder a phenomenon at once so far-reaching, cunning, potent and paradoxical, and so inconsistent with our usual ways of thinking about mind and body, should be regarded by medicine with reserve and suspicion.

The soul-searching that the placebo effect can inspire in medicine is hinted at in an article that appeared a decade ago in the *Journal of Family Practice*.

> Two recent findings highlight the continued controversy over the placebo response. The apparent importance of the placebo response was recently emphasized by the ethical debate over the use of sham surgery control groups in studies of fetal cell brain implants for intractable Parkinson's disease. The need for a sham group and the ethical question of whether exposing subjects to this risk is warranted arises [sic] because subjects receiving the sham procedure typically exhibit marked improvements in their Parkinson's symptoms for up to 6 months and are indistinguishable from patients given the active treatment. This improvement does not seem to be due to either the natural history of the disease or observer bias.[25]

"Controversy," "ethical debate," "sham," "does not seem": the placebo effect appears to pose a profoundly unsettling challenge to medicine. Perhaps if the rituals of daily medical practice nurtured hope and trust—the stuff of the placebo effect—medicine would be able to mobilize the effect with little recourse to controversial procedures; however, the rituals of medicine have frayed, and hope and trust may have frayed with them. I find it suggestive that most Americans reportedly trust their doctor but not doctors in general,[26] which mirrors the divided sentiments of voters who distrust politicians and yet re-elect their incumbent with regularity; politics itself being the arena of controversy, debate and suspicion par excellence.

If, as most informed commentators agree, the placebo effect was once essential to the practice of medicine, its principal vehicle was the very rite of ministering to the patient. The sense of being treated, of receiving care, nourishes the placebo effect, but in order to gain this sense the patient has to be heard, not just processed. With the pressures now bearing on the physician—especially the need to see patients speedily, one after the other—some element of the rite of medicine is sacrificed even as tools and drugs of unprecedented efficacy enter medicine's arsenal. Writes Edward Shorter in a social history tracing the strained relation between patient and healer, "It is, to our postmodern minds, quite incredible that [three-quarters of a century ago] patients expected the doctor to call *virtually every day*"—three or four days successively for the mumps, five days for a nervous condition, and so on.[27] And to call in this context means to call upon. Doctors no longer call upon their patients at all.

Compared to the postmodern physician for whom a call means a phone and time is a commodity in short supply, the attentive physician of the 1920s or 1930s had little power to treat and cure. Hence the use of bromides. Allowing patients to tell their story and hearing them out was itself a sort of bromide, which is not to say that this rite was without therapeutic effect. On the contrary, it is probable that many complaints were alleviated by the release of telling and the consolation of being heard by a gentleman of science, especially if they were nonspecific to begin with. "Suggestion," concludes Shorter, "plays an enormous role in the practice of medicine, even though neither doctors nor patients like to admit it. What interests me is the declining ability of doctors today to cure by suggestion," declining if only because they no longer have either the luxury or the inclination to take the patient's history and devote time to the passivity of listening—to being patient themselves. "Eleven minutes may be enough to make an organic diagnosis and write a prescription, but are they enough to heal?"[28]

Even as physicians at one time helped patients by the rite of attending to them, they or others also played deceptively on the placebo effect by administering "medications" known to them to be useless, from distilled water to sugar pills. A notably cynical account of this practice was given by Louis Lasagna in 1955:

Certain primitive maneuvers are necessary to insure the success of this pharmaceutical charade. First, the patient must be kept unaware of this deceit [a principle now under challenge]. A good start is usually made by the writing of the prescription. The well-known illegibility of scripts frequently makes it impossible for the curious patient even to guess at the nature of the medicament. . . . [However] names such as ammoniated tincture of valerian can safely be revealed to the patient without upsetting the psychological applecart.[29]

An open professional secret, this "charade" was never intended to stand up to the light of public examination, and when subjected to such scrutiny a generation ago it very soon came to appear indefensible. Sissela Bok's historic article questioning "The Ethics of Giving Placebos," published in the *Scientific American* in 1974, opens by telling of a number of

> Mexican-American women who applied to a family-planning clinic for contraceptives. Some of them were given oral contraceptives and others were given placebos, or dummy pills that looked like the real thing. Without knowing it the women were involved in an investigation of the side effects of various contraceptive pills. Those who were given placebos suffered from a predictable side effect: 10 of them became pregnant. Needless to say, the physician in charge did not assume financial responsibility for the babies. Nor did he indicate any concern about having bypassed the "informed consent" that is required in ethical experiments with human beings.[30]

In the most infamous medical study in American history, black field workers in Macon Country, Alabama, afflicted with syphilis (known to them as "bad blood") were given a charade of medical care while in fact the authorities withheld available treatments, eventually including penicillin, in order to follow the progress of the disease right to the autopsy table. Launched in the 1930s and known in medical circles if not to the world at large, the Tuskegee Syphilis Experiment continued of its own momentum for decades until it burst into public notice in 1972, two years before Sissela Bok's article. Immediately notorious, the experiment helps explain the sort of prohibitive disrepute that now surrounds the practice of deceiving patients with sham treatments.

Research into the placebo effect as opposed to the use of placebos as mere controls has intensified markedly in recent years, with each new confirmation of its power and scope leaving practicing doctors right about where they were, however. Given that "the ordering of diagnostic tests appears to improve patient satisfaction and well-being,"[31] should doctors then order superfluous tests to make patients happy? Given that "when the clinician stated positive outcome expectancies as opposed to cautious or skeptical expectancies, most studies found improvement in patient self-reports of reduced anxiety, pain, and distress,"[32] should doctors put on the smile of paternalistic benevolence as their predecessors are now reproached for doing? With attention turning to the physiological mechanisms by which placebos reduce pain (one of their best-attested effects), should doctors go ahead and prescribe sham drugs, or perhaps actual drugs at placebo levels? Considering that a good deal of research into the placebo effect depends on deceptions and infractions of informed consent that would be inexcusable in medical practice,[33] it only stands to reason that this research does not translate well into practice. So dubious both legally and morally are many medical applications of the placebo effect that a principled doctor might well want nothing to do with placebos despite the rising interest in them. The term itself is one of ill repute; hence the proposal to replace it with something more fragrant, like "remembered wellness."[34] It is significant that one of the last strongholds of placebo medicine—the over-prescription of antibiotics, probably to appease demanding patients—has come under heavy attack, though more for reasons of public health than ethics. Interviewed doctors who prescribe unnecessary antibiotics "are aware of the problems of their behaviour in such situations, but the word placebo does not come up."[35]

⇌

Although placebos have fallen from favor in scientific medicine[36] such that their only official place is in clinical trials designed to account for their own confounding effect, nevertheless there remains a market for them. A few years ago it was reported that in their disenchantment with institutional medicine Americans spend some $27 billion annually on alternative forms of it, such as herbal remedies, of whose efficacy "little, if any" evidence exists.[37] But so does psychotherapy offer a livelier experience of the placebo effect than is available in medicine. "Modern patients lose the catharsis that only the 'listening healer' can give."[38] In retrospect it appears that the physician who once treated mental disorders under the guise of treating bodily complaints—humoring the patient with sham prescriptions—has given way to the therapist who treats mental disorders openly but with implicit reliance on the placebo effect. If the doctor's authority once charged his words with suggestive power, now that authoritarianism has gone out of fashion the transactional style of the psychotherapist serves effectively as a conductor of the placebo effect. It seems naïve to assume that a response as powerful, ambiguous, and deeply rooted in history as the placebo effect could be driven out of existence, or confined to unofficial practices or countercultural channels, by the changed conditions of postmodern medicine.

We deplore the dehumanization of medicine, especially the concentration on body parts to the exclusion of the whole person. Psychology takes the person as its mandate. Where patients were once attended by physicians, we now look to the psychologist to attend to us, to listen; the figure of the psychologist listening wisely, concentrating, belongs to conventional lore in its own right. If the doctor takes our history perfunctorily, psychotherapy enables us not only to present our history but to reflect on it, and if the doctor takes care of us but does not particularly care about us, the therapist appears to do both. Placebo benefits that once flowed through the rite of the patient's meeting with "an interested, sympathetic adviser"[39]—and the first to use the term "placebo" in its modern sense, Haygarth's professor of chemistry William Cullen, thought of the physician as just this[40]—have thus passed to the psychologist's office. What is the persona of the therapist if not an interested, sympathetic advisor? As Edward Shorter argues, at one time seeing a doctor for an unspecific complaint could genuinely help the patient, provided

1. The doctor showed an active interest in the patient.
2. The patient had an opportunity to tell his or her story in a leisurely, unhurried way.[41]

Today a patient searching for these good things knows exactly where to find them. When *Consumer Reports* polled readers in 1994 about their experience over the past three years with providers of mental health services including family doctors, psychologists and psychiatrists, a thousand respondents had seen their doctor for an emotional problem and three times that number a mental health professional. Of those who saw their doctor, "significant" numbers were dissatisfied.[42]

That many of the ailments for which patients seek out their doctor remain nonspecific and possibly psychogenic to begin only makes these patients better candidates for a psychological treatment. The common complaint that doctors are too rushed finds its cure, likewise, in the therapist's confessional. When doctors with the exception of psychiatrists could or would not listen by the hour, therapists—sometimes popularly confused with medical doctors—offered to do just this. (Who can imagine a medical doctor meeting with a patient, say, a dozen times, fifty minutes each session, over sixteen weeks?) Even as medicine became more powerful but less personal, psychology surged in popularity, quite as if it had assumed the functions of listening, advising, and comforting defaulted by medicine. By the turn of the twenty-first century there were some 50,000 clinical psychologists among a quarter million psychotherapists in the United States practicing untold varieties of treatment—possibly hundreds, some of which, according to a leading researcher of the placebo effect, in fact border on magic.[43]

From 1979 (five years after Sissela Bok's exposé) to 2007 there were few studies of the use of placebos in American medical practice,[44] an indicator of how touchy or in fact untouchable the subject had become. Good information is still hard to come by, but if, as some think, placebos are most likely to be used to pacify demanding patients who threaten to take up too much time, this in itself would illustrate the acceleration of medicine that has sent care-seeking patients elsewhere. Not only does psychotherapy dispose of placebo effects that are less available to medicine as it becomes increasingly technological and preoccupied with body parts, and increasingly pressed, but the sort of factors deterring the medical use of placebos have no equivalent in psychotherapy. The therapist does not look back to chilling precedents of deceit—men with syphilis treated with aspirin, women seeking contraception and receiving dummy pills instead. Whereas a doctor who prescribes a placebo "may feel a little guilty" nowadays[45] or salve a wounded conscience by informing patients that they *may* receive a placebo,[46] a therapist can proffer comforting but empty words or indeed comforting fictions—for "false interpretations and insights may be just as plausible and credible as veridical interpretations and insights; perhaps even more so"[47]—without necessarily having a sense of offering a placebo at all. Presuming the truth both of his or her theoretical models and of the case histories fitted to them,[48] the therapist could not be farther from a doctor who prescribes a sham treatment. The very freedom to offer placebos and the lack of both cautionary precedents and epistemological checks, all in a setting strongly, indeed uniquely conducive to suggestion, leave the field wide open for the placebo effect.

Some psychologists, while maintaining that psychotherapy does not come under the medical model of diagnosis and treatment, nevertheless do not wish to be associated with the placebo effect. Others have no such aversion, and to the allegation that they cultivate the placebo effect, might say,

"So what?" So say Gerald Koocher and Patricia Keith-Spiegel in their influential *Ethics in Psychology and the Mental Health Professions*:

> Research has . . . taught us that a powerful placebo effect exists with respect to psychotherapy, meaning that good evidence demonstrates that seemingly inert "agents" or "treatments" may prove to have psychotherapeutic benefits. . . . From the client's viewpoint, it may matter little whether positive changes or perceived improvements result from newly acquired insights, a caring relationship, restructured cognitions, modified behaviors, abandoned irrational beliefs, expectancies, *or* a placebo effect. . . . If the client improves as a result of the therapist's placebo value, so much the better.[49]

What if perceived improvements should be only that, perceived? What if new beliefs have rubbed off on the client—what if they themselves should be artifacts of the placebo effect? (After all, the model of therapy that defines interpretations as "means of transmitting the therapist's conceptual framework"[50] practically calls for beliefs to rub off on the client.) What does it mean to say that one's insights are products of suggestion? Such questions are simply passed over. If psychotherapy harnesses the placebo effect, as reflection suggests and the medical as well as psychological literature tends to confirm, this is an occasion for concern, not complacency or indifference, because the placebo effect will recommend false memories as well as true ones, poor as well as good advice, and fanciful as well as sound suppositions. Somehow one of the most ethically troubling things to be said about psychotherapy, that it plays on the placebo effect, is granted in a work on Ethics in Psychology as though it were not troubling at all.

Underlying the acceptance of psychotherapy as an alternative to medicine is its exploitation of the placebo effect—a resource deeply part of the history of healing—without medicine's inhibitions and impediments, as in the passage just cited. The American Medical Association Code of Ethics regulates the use of placebos in clinical practice, permitting their administration only with the patient's consent, a protocol most will find strange, while the corresponding American Psychological Association document makes no mention of placebos at all. Debate over the use of placebos in medicine is also far more robust than debate over their use in psychotherapy.[51] After laying out the risks of the use of placebos in medicine and likening psychotherapy itself to a placebo, the authors of *Persuasion and Healing* do *not* lay out the risks of exploiting the placebo effect in the "powerful influencing situation" of psychotherapy.[52] It seems psychotherapy is an innocent way to harness the placebo effect. The claim that the placebo effect has freer scope in psychotherapy than in medicine finds support, too, in a paper on "The Placebo Response" urging doctors to make the most of that resource by becoming, in effect, therapists themselves. Doctors are exhorted not only to take time to listen to the patient (among other unexceptionable proposals) but, when no bodily ailment can be detected, to do story-work with the patient and to say things like "Between now and the next visit, see if you can discover things that you can do, on at least some days, to make you feel more in control" or "Do you think, now that you have done such a good job of finding the thing that works, that you might think of another?"—all the while taking care to praise the somatizing patient and "to stifle the advice-giving urge."[53] A physician who gets drawn into this sort of dyad has at some point abandoned medicine in favor of psychotherapy.

But the psychotherapist who supports and encourages does not think of this activity as a cultivation of the placebo effect. The former president of the American Psychological Association cited above as denying that the standards of medical research apply to psychology has written a volume entitled *Authentic Happiness: Using the New Positive Psychology to Realize Your Potential for Lasting Fulfillment.* No doubt the author believes that "potential for lasting fulfillment" refers to some actually existing entity analogous to a seed, but the notion that a second, truer, happier, more authentic self resides within waiting to be activated (the psychic equivalent of our celebrated innate capacity for healing) is plainly more fairy-tale than finding. The patient who goes on a search for this mysterious inhabitant is doing story-work indeed.

⊜

It is reported that with the transformation of the hospital at Bath into a renowned center for the study of rheumatology, the once-famous spa went into decline. "Paradoxically, Dr. [George] Kersley [a mover in this transformation] is now one of the most vigorous campaigners to re-establish the city as a spa, lamenting that perhaps he and his colleagues in the heady days of the 1950s had 'knocked the spook out of the waters' too thoroughly, forgetting the phenomenal effect of mind over matter when they insisted on complete scientific appraisal of all treatment."[54] Having exposed the placebo effect as a paternalistic sham and a trick of expectation—having subjected it to rigorous suspicion by controlling for it in clinical trials—medicine itself has knocked the spook out of the waters only too well.[55] But the spook has not been slain, only displaced. If human healing until recently has been a tale of the placebo effect, by the same token it is so deeply embedded in our history that it seems vain to expect it to vanish from the practices of healing even if relics, charms, and waters have lost their magic. The effort to rein in the placebo effect as medicine has done may leave patients disappointed and inspire their search for a fuller enjoyment of its benefits. Arguably, psychotherapy—a fertile field for the placebo effect—offers just this prospect.

As paternalistic medicine came under criticism around the time of Sissela Bok's exposé of the abuse of placebos, talk turned toward partnership between patient and healer. Far more than medicine, psychotherapy is premised on partnership, and insofar as the client is an active party, his or her investment in therapy's course and conclusions is apt to be greater. But what if the appeal of the "story" constructed jointly by therapist and client should reside in its way of satisfying narrative conventions? The archetypal story-teller in our tradition—Odysseus—is a master at weaving yarns that sound compellingly true because of their twists and turns, but are pure fiction. Austen and Tolstoy, among others, wrote novels about the false attractions of stories.[56] A story-line whose familiarity gratifies expectations would seem a natural vehicle for a placebo effect largely dependent on expectation. If it is true that "to be effective, interpretations, the primary means of transmitting the therapist's conceptual framework, need not be correct, only plausible,"[57] the possibility that the conclusions of therapy may persuade precisely because they are familiar—conventional—is certainly in play. False memories of sexual abuse retrieved in the heyday of the Recovered Memory movement may have rung true not only because a history of sexual abuse theoretically explained the patient's symptoms but because tales of sexually abused children had become a genre, thus lending a semblance of plausibility to the memories in question.[58]

Just as the possibility of ill-founded therapeutic insights and interpretations is immediate, not remote, so the issues at stake are anything but academic. If "any therapist or healer who can establish a comforting relationship with a patient by taking the time to listen, regardless of any theory behind what he or she does, will lighten the patient's perception of the problem,"[59] then by the same token the placebo effect will work to recommend potentially anything the therapist may suggest, imply or advise, whether well-founded or not. The most philosophically rigorous study of psychotherapy yet written finds the talking therapies—in particular, but perhaps not exclusively the psychodynamic therapies—extensively contaminated by an epistemological

license that authorizes fictitious theories and spurious insights, the worse because the object of this pseudo-knowledge is our very selves and because the
theories (etc.) are credited as if they were not epistemologically compromised
at all. Writes David Jopling in *Talking Cures and Placebo Effects*,

> There is an . . . ethical dimension to the idea that truth matters. False, bogus,
> or fictional psychodynamic interpretations and insights can be as psycho-
> logically harmful as false memories. Like false memories, they can lead to
> the break-up of families, the dissolution of marriages or partnerships, the
> radical alteration of life plans, the erosion of religious faith, or the morally
> self-serving rewriting of the past. What looks like *bona fide* insight, or self-
> knowledge, or a genuine realization, or a new and more empowering way of
> looking at oneself, may in fact be ethically calamitous.[60]

That truth matters might go without saying, except that a seminal study of
psychotherapy maintains explicitly, and somehow without exciting controversy, that it does not.[61] As long as therapists who engage the power of the
placebo take the position that "the 'truest' [psychotherapeutic] interpretation
would be the one that is most satisfying . . . to the particular person,"[62] or
that "the truth or historical reality of their patients' assertions" is not to be
put in question,[63] or indeed that mental health requires positive illusions (as
in the Pollyanna proposal), the possibility of ethical calamity will remain
a clear and present one. I mentioned above that the healer who believes in
his or her words and deeds, as opposed to merely playacting, is especially
well positioned to exercise suggestive power.[64] Jopling concludes that most
practitioners of the talking cure believe all too much in their own theories
and explanations. He finds among them "little awareness . . . of the epistemic
complexities of psychodynamic insights and interpretations, coupled with
high levels of epistemic confidence and theoretical self-assurance about their
authority."[65] Such practitioners risk abusing the placebo effect because their
belief in their insights and pseudo-insights makes them all the more persuasive and because their play on the placebo's power is bound up with laudable
goals such as "combating demoralization." Suggestion is a dangerous game.

So questionable is the pursuit of insight under the auspices of the placebo effect that Jopling recommends that patients be warned extensively of
the pitfalls awaiting them. In a spirit of transparency they are to be notified of "the role of suggestion, placebo and expectancy effects, evidentiary
contamination, psychodynamic artifacts, common factors, the Barnum effect

[the seeming plausibility of generic personality profiles], and other factors that could interfere with clients trying to acquire self-knowledge, or trying to 'get in touch' with an 'inner' or 'core' or 'authentic' self."[66] One has only to read this formidable disclaimer to see the unlikelihood of its ever being put into effect. It conflicts with the practice of psychotherapy itself, somewhat like the voice in a drug commercial that recites side effects while the images show people playing golf.[67]

Chapter Thirteen

Can Placebos Survive Disclosure?

Open placebos would tap the placebo effect without violating the dictates of transparency.

Besides inducing the side effects of the drugs they stand in for, placebos raise concern on the grounds that they may distract doctors from the practice of sympathetic attention, relieve symptoms while leaving an underlying condition untouched, turn off the alarm function of symptoms themselves, breed dependence.[1] As I have argued, the placebo effect can also validate all manner of "story-work." All in all, however, placebos perform remarkably well—perhaps only too well—considering their composition. This surprising efficacy combined with their potentially broad application in medicine helps account for the surge of research interest in placebos in recent years.

A decade after his landmark paper on "The Powerful Placebo" Henry Beecher published another no less important, "Ethics and Clinical Research," which by documenting and deploring the practice of experimenting on patients without their knowledge helped inaugurate the era of informed consent.[2] Though much experimentation on the placebo effect employs deception under the guise of informed consent,[3] and though we still hear of placebo treatments it is difficult to imagine anyone knowingly consenting to, such as the implantation of a pacemaker switched to Off,[4] in the post-Beecher era most of us believe in transparency. A saline solution that acted like a drug even though it was known to the patient to be saline solution would be a triumph of transparency. If placebos could be administered openly, the wall separating research from clinical practice would come down and medicine could exploit the manifold power of placebos that has been shown in one study after another—a power perhaps even more far-reaching than Beecher supposed. The proposal that therapists inform patients of the placebo content of psychotherapy itself is but one marker of rising enthusiasm for the possibility of open placebos.

But the findings of placebo research may not translate into clinical practice. Although research has shown that the results of sham arthroscopic surgery on the knee may rival actual surgery, I can't see myself seeking out fake surgery for an arthritic knee. Suppose, however, despite everything, that a patient who learned that fake knee surgery seems to work, and is no more dangerous than an injection, did seek it. What ethical surgeon would accommodate? He or she would refuse because sham surgery was intended not as a medical procedure but as a test of the efficacy of performed surgery, and because its success depends on the deception of the study subject. A patient seeking out sham surgery wishes, in effect, to pretend to be deceived.[5]

However, it is sometimes said that subjects don't really need to be deceived in order for placebos to work their wonders. In a notable paper published some thirty years ago Howard Brody and David Waters claim that "Even when patients are informed of the inert nature of the placebo, they may respond positively,"[6] their authority for this contention being a small study conducted in 1965 that employed no control group and was never replicated.[7] More recently David Jopling, investigating the possibility of open placebos, offered the guarded conclusion that "While this has not been the subject of much research, there is some clinical evidence to suggest that patients who are informed that they are receiving saline injections, sugar pills or other placebos sometimes continue to experience measurable objective symptom relief."[8] The evidence referred to turns out to be the same unreplicated 1965 study faulted for "a small patient sample, questionable symptom matches and comorbidity profiles between patients, an overly short treatment course, no wash-out period for potentially confounding psychoactive medications taken by patients, and no control groups (e.g. a no-treatment group)" by Jopling himself.[9] It bears noting that the fifteen "neurotics" who constituted the population of the study in question were specifically told twice that the sugar capsules they were being given helped many others with similar conditions, and that this strong recommendation was "usually repeated again, especially if the patient asked questions concerning the treatment, conveying doubtful attitudes about its possible effectiveness."[10] The invitation to experience the same benefits as others seems to have been more emphatic than the disclosure of the medical nullity of the capsules. In that the others already helped by sugar pills took them in the belief that they were an active medication, the open placebos in this study turn out to be somewhat less than open; a degree of deception was grandfathered into the study protocol. (It is also noteworthy that the study concludes that treatment with sugar capsules "could be viewed as having some affinity to psychotherapy.") While open placebos have received very little experimental validation, the principle that we tend to experience what we believe others do has been confirmed in studies and illustrated in life at large time and again.

Recently the foremost investigator of the neuroscience of the placebo effect reported a study in which athletes treated with morphine in training but placebo on the day of competition responded to the placebo as if to the drug, a carry-over that raises the possibility of achieving "drug-like effects without drugs" in real-world conditions. However, the experiment hinged on a deception. Far from receiving truthful information, placebo groups were told on the last day that they were getting morphine and should expect an increase in pain tolerance—a strong message indeed.[11] Even given our impressive capacity for self-deception, it does not seem credible that athletes in the real world, looking to circumvent doping regulations, would train with morphine the better to fool themselves into mistaking placebo for morphine at a later date. (Writes a medical commentator on the placebo effect, "I doubt that one could give oneself a placebo.")[12] Elsewhere the author cites a body of research showing that expectation strongly influences placebo analgesia, which is to say that placebos engender analgesic effects in good part because we expect them to behave like the active drugs we mistake them for.

Placebos score high in clinical trials of antidepressants. Of interest, therefore, is the postscript of a study of placebo antidepressant in which the subjects were eventually informed that they were in fact in the placebo group. According to the study director,

> At eight weeks . . . you couldn't tell [the treatment and placebo groups] apart in terms of mood ratings. What happened at eight weeks plus a day is a bit different. Some of the placebo responders, when told they were on a placebo, had a deterioration of their mood. In fact, most of them did. Within a month, most of the placebo responders had enough depressive symptoms that they actually ended up on medications.[13]

While this result will disappoint those who believe the placebo effect can survive unblinding, it seems consistent with the body of placebo research. (And with common experience. Now that the mineral waters of Bath are known to have no curative value, people no longer flock to Bath to enjoy their benefits.) If placebos didn't depend on concealment, it is hard to see why their use in research would be so bound up with concealment.[14] Similarly, while placebos in one guise or another are used in clinical practice, only rarely are they prescribed openly. A recent survey of Canadian doctors found that while somewhere over half said that they used placebos now and then, only five percent reported "telling their patients that they were receiving a placebo."[15] Why would so many conceal the placebo, thereby putting themselves in violation of ethical principles written or unwritten,[16] unless its efficacy depended on concealment?

Unless and until it is refuted by robust evidence from replicated experiments, the presumption must be that placebo responses to medications, especially for pain, do hinge on deception. A challenge to this presumption is an unusual study of some 26 children with ADHD reported in 2008. To determine if part—not all—of their medication could be replaced with placebo, researchers offered the children, who were supervised by their parents, certain capsules along with a clear explanation that they contained no drug but might boost the effect of regular medication. Persuaded of the power of placebos but opposed to deception, the researchers explicitly assumed that revealing the placebo in this way would not destroy its efficacy. "We hypothesized that disclosure would not eliminate the placebo effect." Though the assumption was confirmed, the published study contains the following weighty disclaimer: "This pilot study has important limitations, including very short-term treatment outcomes and relatively few subjects. The outcome measures are inherently subjective and the open-label study design introduces the potential for bias. Teachers were the only blinded raters during the study, and the teacher data did not show significant differences in child behaviour among the [experimental] conditions."[17] As in this case, enthusiasm for open placebos tends to run well ahead of the evidence.

A similar ADHD study had the placebo disclosed to children as young as six.[18] Can a child of six understand something as paradoxical and bewildering as an inactive substance that happens to be active? For their part, children from ten to twelve in this study were told,

> This little capsule is a placebo. Placebos have been used a lot in treating people. It is called 'Dose Extender.' As you can see, it is different from Adderall. Dose Extender is something new. It has no drug in it. I can promise you that it won't hurt you at all. It has no real side effects. But it may help you to help yourself. It may work well with your Adderall, kind of like a booster to the dose of Adderall. That's why it's called a Dose Extender. I won't be surprised when I hear from you and your parents and your teachers that you're able to control your ADHD better.[19]

The open placebo turns out to be a highly leading rigmarole. Note too the suppression of the critical fact that the placebos "used a lot in treating people" are not known to them to be placebos.

In another defense of open placebos that comes up short, a group of researchers in 2007 conducted two studies designed to determine whether "learning that pain reduction is the consequence of a placebo treatment reduce[s] responses to subsequent placebo treatments." The second of these experiments "employed repeat sensory testing after participants were informed about their previous placebo response, allowing for assessment of the effects of such knowledge on subsequent placebo responding." At a certain point in this pain study, then, subjects in the placebo-informed group were notified that they had actually received a placebo cream. However, they were then told that they would now receive an active cream, while in truth they received the placebo for a second time. They continued to respond as before to the sham medication. "Interestingly," conclude the authors, "the placebo effect persisted when a second placebo cream was applied even after participants were told that the first cream used in the study was a placebo. Although the strength of that second placebo was slightly reduced, approximately 84% of the original placebo effect remained."[20] But surely this does not establish the efficacy of an open placebo. On the contrary, the subjects were specifically told the second time that they were being given an active medication. They were lied to. They fell for the lie presumably because the researchers made themselves appear honest by confessing that the first administration of the cream was a sham. The experiment in question was a study of the possibility of fooling someone twice.

As this case illustrates, placebo research sometimes converts the very disclosure of information into an act of equivocation.[21] Often, it seems, a placebo is craftily described to induce study subjects to mistake it for an active drug. Thus, for example, in a study investigating the effect, if any, of verbal suggestion on analgesia, a number of patients suffering from irritable bowel syndrome were told, "The agent you have just been given is known to significantly reduce pain in some patients"—the magic words—when in fact they had been given a placebo. This invitation to feel what others feel, which resembles in outline the sales pitch for the Perkins tractor used by experimenters two centuries ago and appeals powerfully to our social nature, was intended specifically to arouse an expectation of pain relief, in contrast to a similar study in which patients were truthfully informed that they "may receive an active pain-reducing mechanism or an inert placebo agent." Though it is highly unlikely that the patients interpreted the pitch to mean they would receive a placebo—after all, it was phrased to create exactly the opposite impression—the authors contend that because the placebo did work in a previous study, their statement wasn't really a lie. On this lawyerly claim they ground the inference that verbal suggestions for pain relief in general "need not be deceptive and thereby ethically problematic."[22]

As in this instance, the principle that we tend to feel what others do, and expect the therapies that help them to help us, crops up in proposals for the ethical use of placebos. A proposal for the use of placebos in clinical practice suggests that doctors offer placebos to depressed patients with cover language like the following: "I do not know why you are depressed—modern medicine does not understand depression very well. It could be that you have a chemical imbalance or it could be due to stress in your life. Trials have found that 60% of patients feel significantly better when they take an antidepressant, so that is what I am prescribing for you."[23] This statement, which logically resembles a sort of private joke, conceals the cardinal fact that the antidepressant being prescribed contains no active ingredient at all: yet another careful equivocation. Also exploiting ambiguity, some doctors now recommend "probiotics" that certainly sound to the lay ear like medications, but for which there is little or no evidence of efficacy. Popular with consumers and believed to be harmless, these commodities fit the profile of a placebo and are probably discussed in doctors' offices with the same craftiness.

Some believe, however, that under certain conditions placebo-prescribers could dispense with disguise. According to a philosopher,

> Were the general efficacy of placebos well accepted, and, in particular, were it well recognized that successful treatment by placebo does not indicate that an illness is merely imaginary or that the patient is of a peculiarly gullible or dependent personality type, there would be no reason for deception in their administration. In those cases where placebos may reasonably be expected to be useful, and where pharmacologically active agents are ineffective or contraindicated, a physician could simply report to a patient that the prescribed agent appears to be pharmacologically inert with respect to his or her disorder, but that *in fact*, it has been shown to be therapeutically effective in other patients suffering from the condition.[24]

The suggested script—"the prescribed agent appears to be pharmacologically inert with respect to . . ."—still has a certain studied obscurity. But why is it that the line that "others have benefited" suggests itself to the defender of open placebos? No doubt because it is the strongest recommendation of an inert treatment that could be given with technical veracity. Addressed to our social nature, it is a tribute to the principle that we model our experiences on the reported, imagined, or presumed experiences of others. Note that the philosopher's argument assumes not the efficacy of placebos per se but public acceptance and recognition of their efficacy, so that candidates for placebo treatment will "simply" be asked to believe the same things that others generally do. Placebo efficacy under these conditions might turn out to be a pyramid scheme, with people investing belief in inert treatments because others do the same, until the entire structure collapses.

Unlike studies in which supposedly open placebos turn out to be cloaked in artful language, a recent study of placebo treatment of irritable bowel patients saw the treatment group given pills described as being inactive "like sugar pills" and explicitly labeled as placebo. After three weeks, 59% of patients treated with the placebo reported adequate relief as compared to 35% of the untreated control group—a finding qualified by a number of limitations laid out in the report of the study itself. Like Natasha in *War and Peace* who takes solace in the ritual of regular dosing ("though she declared that no medicine would cure her and that it was all nonsense"), the study subjects may have responded to the medication ritual that was withheld from the untreated group. Unlike Natasha, though, the study subjects had reason to believe their pills were effective, whatever they were labeled as. Potential subjects were told that half the study population would receive inert pills "which had been shown to have self-healing properties" [sic] and those entering the study that "placebo pills . . . have been shown in rigorous clinical testing to produce significant mind-body self-healing processes," so that the group that proved so responsive to placebo received a double dose of the message that others are known to benefit from the placebo pills and they can expect to as well.[25] Not that placebos actually "heal"—that is, cure—IBS; at best they alleviate symptoms.

If open, rather than deceptive, placebos had already been shown in rigorous testing to be therapeutic, there would have been no need for the experiment; it would have been redundant. The experiment does not establish that pills frankly described as containing no medication can have beneficial results. At best, it opens the possibility that placebos revealed as such, but hyped as having some kind of "healing" power and extolled as being of proven benefit to others—who in all likelihood believed them to be active medications—may have beneficial results. In effect, then, the study put into operation the principle that if only people generally accepted the efficacy of placebos and the legitimacy of using them, they would no longer require deception. Ideologically, it is in the tradition of More's Utopia, where medicine is all but unnecessary, people think alike, and everyone lives "in the full view of all,"[26] without the need or even possibility of concealment. Just as the Utopians honor medicine highly but are less in need of it than any other people (so More tells us), the use of open placebos will enable "healing" without the use of medicine.

Some seem to believe, similarly, that by making the patient's condition comprehensible and pointing to something to remedy it, the very act of diagnosis constitutes a treatment. Diagnosis

> is medicine's way of explaining symptoms. The extent to which the explanation will satisfy the patient will depend on the extent to which he shares the physician's presuppositions about what sorts of things cause and contribute to disease and healing. Secondly, the diagnosis is often a crucial factor in encouraging the expressions of caring and support from family and friends. Before the patient's changed behavior has been given the interpretive label of a diagnosis, others may be uncertain as to how to react to him or her; but once the physician as the authority figure has legitimized the behavior with a diagnosis, the patient has "a mantle of distress that society will accept." Thirdly, the ability to give something a name implies the ability to gain control over it. This is true both in magical belief systems, where words and names have special powers in and of themselves, and in scientific belief systems . . . [27]

What's in a name? A great deal, evidently. Note, however, that the authors do not concern themselves with the possibility of a mistaken diagnosis, evidently because the act of diagnosis per se is therapeutic, whether medically accurate or not. But surely something is wrong with a medical argument indifferent to the possibility of a diagnosis itself being wrong.

In view of the risks of offering diagnosis per se as a treatment, and the ethical traps of using placebos in ways that seem open but perhaps aren't, or of disguising them as active medications (as in the prescription of drugs at sub-operative doses), the best and least controversial way to exploit the placebo effect in medicine is surely the humane, attentive practice of medicine itself. By consensus, after all, the manner and behavior of doctors contribute richly to the "power of context" that frames the placebo effect. A study of the differential components of the placebo effect among patients with irritable bowel syndrome found the most potent contributor to be the quality of the relation between physician and patient, as measured not only by time spent with the patient but the projection of both confidence and sympathetic concern.[28] Perhaps if doctors were less hurried and more attentive to care as well as cure, interest in placebos as short cuts to health would not be running so high. If it is really nothing but a "reaffirmation of man's essential tie with his sociocultural nexus" as some theorize,[29] then the placebo pill, the subminimal dose, the probiotic, is secondary to the actual bond between patient and doctor anyway. If the power of context activates the pill—and some now refer to the placebo effect as a context effect—perhaps in some cases we can forego the pill and make do with the context.

In the course of the second of the ADHD studies cited here, subjects formed "strong relationships" with the medical team, while parents became "more attentive" observers of their children.[30] Given these circumstances, it is entirely possible that social bonds are the really operative factors in the study—that the power of context is speaking through the Dose Extender. The cardinal health benefits to come to us through social channels actually go quite beyond the effects imputed to a pill. As noted, in a number of studies social connection—marriage, bonds with extended family, and other forms of affiliation—correlates inversely with the risk of mortality itself.

Belonging to a family is a more robust form of membership than participating in a trend or even movement, though "the negative or conflictive aspects of social relationships need also to be considered, since they may be detrimental to the maintenance of health,"[31] as both medical literature and the literature of the imagination well know. In any case, it is impossible to be married without knowing it. There can't be a clinical trial in which one group believes itself married but actually isn't, while another is actually married but is led to believe otherwise. The question of deception does not and cannot enter into the matter of social connection. In More's ideal commonwealth there is no need to resort to medical trickery because the social institutions of a closely integrated people serve as the guarantors of health.

In keeping with the principle that social bonds enhance health, and more specifically with the theory that "Positive emotions and ideas can help to heal the body through the powerful placebo effect"[32] (a theory that may or may not be utopian itself), studies have investigated whether support groups improve survival for breast-cancer patients. At first it appeared they might—a result doubly to be welcomed, first of course for its own sake, secondly because this mobilization of the placebo effect was achieved without deception of any kind. The tested treatment consisted of a series of sessions "designed to build new bonds of social support, encourage expression of emotion, deal with fears of dying and death, help restructure life priorities, improve communication with family members and healthcare professionals, and enhance control of pain and anxiety."[33] Naturally this "supportive-expressive group therapy" was not advertised as something else; disguise was neither possible nor necessary. Later, however, it was found that the survival benefit associated with the therapy could not be replicated[34]—a reminder that the benefits of social bonds have their limits, that it is one thing to inhibit suicide, as Durkheim found, and another to arrest cancer, and that transparency, however desirable ethically, may have no particular therapeutic merit.[35]

Chapter Fourteen

Suicide in Dystopia: "Howl"

The Beat anthem both mourns and celebrates suicide—an epidemic of it.

A salient peculiarity of the psychogenic ills of our time is that many generated or were reinforced by movements—chronic fatigue syndrome, for example, giving rise to an Action Campaign in Britain,[1] and claims of recovered memories being advanced by a corps of believers with its own leaders, foot-soldiers, literature, doctrines, rhetoric, and statistics. Participants in such crusades might be considered as an extended support group, united by the feeling of shared experience. The social electricity flowing through the movement charges the group's language with a special intensity and may serve to relieve the demoralization to which all are potentially subject in isolation. "Patients who participate in an advocacy movement often lose their complaints."[2] Ideas running counter to mainstream thinking, which might have the depressing effect of alienating others when held in isolation, can serve as a banner when held in common. Where Benjamin Perkins insisted that his tractor worked despite the ridicule and censure visited on it by much of the medical establishment, the partisans of, say, chronic fatigue syndrome may argue that medicine's inability to find a cause for it reflects on medicine itself, not on the experience of the patient.

Some epidemics of recent occurrence—among them, multiplying claims of alien abduction and satanic ritual abuse—have no biomedical component at all, being products of collective fantasy. Elaine Showalter designates these and others "hysterical epidemics."[3] But if the height of hysteria is to affirm hysteria itself as a form of superior, virtually prophetic insight, then that peak was scaled well before the outbreaks of the 1990s.

"I saw the best minds of my generation destroyed by madness, starving hysterical naked": recently I came across these words—the opening line of "Howl" (1956)—printed on a tee shirt. The garment's designer chose well, for the line not only proclaims the Beat revolt, it is "Howl" itself, establishing the poem's voice as well as the poet's claim to a witness's authority and vatic insight, affirming hysteria as an attribute of superior minds, and laying down like a first principle the theme of martyrdom or self-martyrdom enforced again and again in the ensuing lines. The entire manifesto is dominated by the most elementary device of rhetoric: repetition. Where the Declaration of Independence enumerates a long train of abuses and injuries, the counter-declaration of "Howl" chants the sufferings and exploits of the Beat elect. Fittingly, the spirit of the ode is the opposite of Jeffersonian reason: Blakean excess. Also in the tradition of Blake and his insistence on mind-body unity,[4] Ginsberg portrays the best "minds" as performing bodily actions, just as he shows them afflicted with a starvation and a nakedness that might or might not be physical.

Animated as they are by communal passions, social movements excite our propensity to feel what others feel, and the catalogue of exploits following the opening line of "Howl" suggests that the best are possessed by identical passions as if they belonged, in fact, to a single movement. Their exploits seem like so many ways of burning themselves up. As the poet saw the best destroyed, so he raises before our mind's eye the images not only of madness and destitution but suicide, and not one or two suicides but a multitude—a kind of epidemic. The best are something like members of a family "detrimental to the maintenance of health."

The heroes of "Howl" pursue their own undoing as if their defiance of the dictates others live by, right down to the elementary dictate of self-preservation, confirmed their membership in the elect. Some "create great suicidal dramas on the apartment cliff-banks of the Hudson," some utter "harlequin speech of suicide" on the steps of the madhouse, others—flirting with the act rather than committing it—"cut their wrists three times successively unsuccessfully," while still others hunt for "an angry fix," destroying themselves over time. Suicide is everywhere. Mourned and celebrated are those who jump from common roofs and ledges, from the Empire State Building and the Brooklyn Bridge ("this actually happened"), disappear into volcanoes only to emerge in California, drink turpentine, manage to fall from subway windows, leap into filthy rivers, find themselves under trucks and taxis, and generally flame out, their way of perishing a sign that they bear a truth too great for this world. Why is there so much self-destruction in "Howl"? Why this festival or this plague? The answer Ginsberg's dream-vision seems to return is that these rebels, death's volunteers, are martyrs one and all of their society. But martyrs choose death, and it may be that the reason Ginsberg's saints choose death is that self-destruction has become for them a kind of social statement. While suicide marks the poem's heroes as defeated, beaten, Beat, it also proclaims them *as* heroes—the best of their generation. As melancholy became a fashion in certain circles of Shakespeare's England because it was thought to betoken genius,[5] the beaten of "Howl" show themselves geniuses of despair.

⊟

The citizens of More's Utopia enjoy and celebrate their wellbeing in common—a strong fictive illustration of the principle that social integration supports health. The citizens, or at least the first citizens, of hell on earth celebrate their despair by destroying themselves.

Thus, while suicide is closely, even ritually regulated in Utopia and unauthorized suicide subject to the most severe censure, suicide runs loose among the rebels of "Howl" like a mode of expression. One act of suicide seems to inspire another. Though the poem decries "the iron regiments of fashion," the self-destroyers who are its heroes behave like people caught up in a frenzy of mimesis, which is to say a fashion of their own, acting out the same revolt in the same way. The protest against convention readily becomes a convention, and in the hell of "Howl" self-destruction paradoxically becomes a way of life.

Nothing could be farther from the decorum of Utopian life as portrayed by More than the practice of staging one's own end in the most provocative and sensational manner possible, as if exulting in the disgrace long attached to suicide. Utopia is a static society; to remain healthy and in good working order it need only preserve its way of life as originally laid down by King Utopus some 1760 years before. It does not need shaking up. By ending their lives so expressively, the suicides of "Howl" do just that. Beyond putting an end to their suffering, their death sends waves of energy through the world around them, all the more if the act is a flamboyantly public one. In accordance with dialectical style, then, the suffering of the suicides is also exaltation, their disgrace is glory, and their end fruitful.[6] A. Alvarez's study of suicide, *The Savage God*, bears an epigraph from Bakunin: "The passion for destruction is also a creative passion." The saints of "Howl" accordingly *"create* great suicidal dramas," scaling the heights of experience on the cliffs of the Hudson, the Brooklyn Bridge, and other prominences.

One important way in which suicide reveals itself in "Howl" as more than just an act of futility is by setting off other acts of the same kind—contributing to a tide or current highly charged with social potential. One act of self-destruction seems to beget another and live on in another, which may be why all of the dead are elegized as if victorious in defeat itself. Through suicide they achieve transcendence. If the martyr vanquishes the oppressor, the legion of martyrs in "Howl" not only vanquish but survive in song. As suggested by the poem's very form as a loose epic catalogue (and according to Ginsberg the essence of "Howl" was its form), each suicide is moved by a common spirit, each feels what the others feel, all are lit up with glory. Self-destruction has become a Dionysian movement, a source of social energy, inspiring those swept up in it and holding out the promise of renewal to society at large, if only the promise were understood. The poet chants like one mesmerized by the vision of such creative destruction.

Not only does "Howl" present itself as an act of iconoclasm, but the idol it smashes is named—sung—in the text itself:

> Moloch! Moloch! Nightmare of Moloch! Moloch the loveless! Mental Moloch! Moloch the heavy judger of men!

But who is Moloch?

According to another visionary poet, Moloch is the first of the idols that overran the earth. The catalogue of idols in Book One of *Paradise Lost* begins with this false god, "horrid king besmeared with blood / Of human sacrifice, and parents' tears." For Milton the enumeration of idols requires nothing less than a lengthy catalogue because idolatry is so subtly pernicious that it has spread everywhere. For Ginsberg, on the other hand,[7] the enumeration of human sacrifices in the form of suicide requires a sort of epic catalogue because suicide is so creative an act that each suicide calls forth others; it sows the dragon's teeth from which spring warriors against Moloch.

Maybe there is such a thing as an outbreak of suicide inspired by a common fantasy. It is said that Goethe's *Sorrows of Young Werther* gave rise not only to certain fashions of dress but a fashion for suicide itself. De Vigny's play *Chatterton* reportedly doubled the suicide rate of France in the 1830s.[8] Such was the influence of the cult of sensibility that it extended as far as

Russia, where, also in the 1830s, a coterie centered on the young Bakunin—"The passion for destruction is also a creative passion"—dedicated itself to a romantic ideal higher than existence itself, one in whose name they professed themselves "ready at any moment to . . . perish."[9] By perishing voluntarily, Ginsberg's heroes—both blessed and cursed—not only escape the jailhouse of the world at large but prove themselves members of the spiritual elite and set an example for others who may aspire to the same company. The poem portrays as if from within a social epidemic in which a shared mythology produces actual, even fateful effects.

⇄

In *Paradise Lost* the angels fall "by their own suggestion" (3.129), which has been interpreted to mean "by a kind of spontaneous mass contagion."[10] Suicide seems to work in the same way among the fallen "angelheaded hipsters" of "Howl."

In a more prosaic case, in 1998 over 150 persons at a Tennessee high school reported being sickened by toxic fumes after a teacher detected the scent of petroleum in her classroom and, along with a number of students, was taken by ambulance to the local hospital. Investigation revealed no source of the fumes. In a report on this classic instance of the nocebo effect published in the *New England Journal of Medicine* it was noted that those who observed or knew of others who fell ill were more likely to fall ill themselves, and that psychogenic incidents of this kind occur in "a group of people with shared beliefs" about the source of the toxic agent.[11] Ginsberg for his part believed his society was being poisoned by a military-industrial smog,[12] and in "Howl" celebrates the sufferings of an elite visited with the same experience of hysteria. In fact the poem "mythologize[s] Ginsberg's own life as well as the lives of a small group of friends 'destroyed by madness,'"[13] with the act of suicide effectively completing the destruction caused by the toxin in the air.[14] The temptation of suicide was familiar not only to Ginsberg, whose mother believed there was a conspiracy to destroy her and attempted suicide, but to his circle, for by the time he wrote "Howl,"

> Ginsberg knew at least half a dozen individuals who had either attempted or committed suicide. He had listened to Kerouac talk about suicide for years, and he'd watched Burroughs's self-destructiveness, too. There were his own ongoing suicidal impulses and there was suicidal American society at large—American military and material self-destructiveness.[15]

"Howl" presents itself as a hymn to heroes defeated, beaten, Beat. However, it can also be read as the document of a psychogenic outbreak, a spontaneous mass contagion, in which a common fantasy that the world is causing their destruction takes possession of a group of people one after another.

In recent decades outbreaks of fantasy have proven all too real. Surveying delusions like ritual satanic abuse that spread in the 1990s by a kind of semi-spontaneous mass contagion, Elaine Showalter identifies them as virulent forms of collective hysteria. "Individual hysterias connect with modern social movements to produce psychological epidemics."[16] Something like this mutation is on exhibit in "Howl," whose visions of Moloch—the all-powerful complex that is also the god of human sacrifice—are no less extravagant than the fantasies of organized evil cited by Showalter. Setting his insights to the drumming of the Beat movement, the poet offers his own hysteria as the spirit of the age.

Chapter Fifteen

The Prostate Cancer Epidemic—
What Spawned It?

Enthusiasm for prostate cancer screening is as much socially as medically driven.

In contrast to a psychogenic outbreak, some epidemics of our time have taken place under the auspices and authority of medicine itself. Depression rose to epidemic levels when a strong trend toward diagnostic inflation, abetted by the mass marketing of antidepressants, defined millions of people with a few symptoms consistent with depression as sufferers from the disorder per se. Earlier studies pointing to a depressed population of disturbing magnitude led to efforts to screen for depression, which in turn led—not surprisingly—to the discovery of still more depression, which was validated by a mushrooming medical literature. Recently it was reported that "the number of articles concerned with depression is now far higher than any other psychiatric diagnosis and has grown far more rapidly than the general growth of psychiatric research publications."[1]

We now consider an epidemic of prostate cancer, more or less concurrent with the depression epidemic[2] and similar both in that it too is driven by screening and overdiagnosis and it too has an ambiguous medical basis (by analogy with symptoms of depression shading off into normality). In both cases the public is alerted to the existence of a disorder said to be all at once prevalent, dangerous, and potentially invisible, but which, despite its invisibility, can in theory be identified before it advances to something worse. In both cases the seemingly irresistible argument is made that the identified condition must be treated early lest it progress. Not only did these medical movements use a similar mobilizing rhetoric, but each reaped a multiplier effect by appealing to a spirit of medical activism that was much in the air at the moment, and that powered the other. With prostate cancer, however, the treatment is very often worse than the disease, which may qualify the epidemic as a nocebo phenomenon—one where psychosocial causes generate adverse effects. In 2007 it was estimated that over a million American men "who, but for [screening], would have lived as long without being a cancer patient," had already been treated for the disease,[3] with all the side effects, including impotence, that treatment commonly entails.

☰

Both placebo research and the literature of the imagination suggest that our own experience can be unreliable and that imagination can creep into a sensation even as seemingly indubitable as pleasure or pain. Does Eve really experience intellectual pleasure upon eating the apple? Is it really the case that a brand name "can make aspirin work better," as an interpreter of the placebo effect claims?[4] For that matter, are people truly sedated when they take blue pills, regardless of the pills' content? Doctors too, however, can be misled by experience and perception. The case has been made that in their enthusiasm for prostate-cancer screening, doctors resemble pilots who mistakenly rely on their senses when flying in confusing conditions.

> Intuition—particularly in the field of cancer screening—can easily lead physicians astray. This is because there are powerful biases associated with early detection strategies that can fool even the most careful observer. An analogy to flying can be made here—in common situations (such as when flying in cold front clouds, or where haze meets the water line), pilots can be led astray by their senses, and this is the reason for the existence of instrument flying. Pilots learn not to rely upon their own observations in certain scenarios, which can deceive them.[5]

While the promotion of mass screening for prostate cancer at the behest of clinical experience and intuition puts that practice into conflict with evidence-based medicine, the principle that screening saves lives—the assumption that launched the prostate-cancer screening revolution, and therefore precedes the evidence rather than following from it—resonates so strongly with the public's own intuition that millions of men get tested regardless of the ensuing harms, a vote of confidence that in turn seems to reduce the arguments against the test to skeptical caviling. Even though medicine well knows that most prostate cancer is indolent and that autopsies on men who died of other causes commonly find such cancer (and that any number of the doctors who order and evaluate tests for prostate cancer harbor dormant disease themselves), many of the screened who are discovered to have prostate cancer and are then treated for it believe passionately and unshakably that screening saved their life—another demonstration of the unreliability of experience.[6] If everyone treated for screening-detected prostate cancer had his life saved, the disease's mortality rate would be far higher than it was before screening.[7]

Introduced in the United States some twenty years ago, screening for prostate cancer has caught on to the point that somewhere over half the eligible population now has PSA (prostate-specific antigen) testing. But while we might have expected that such an investment in a test would be underwritten by strong evidence of its medical value, PSA testing has followed an unusual path. Even as the test acquired millions of takers and established itself firmly in prostate-cancer medicine, its very basis came into question. Indeed, no sooner were men lining up to have blood drawn in the name of prostate-cancer "awareness" than doubts about the wisdom of doing so were expressed in the medical literature. The flaws of PSA testing—in particular, its speculative benefits but probable harms—were flagged virtually from the beginning,[8] although cautions against the over-eager pursuit of early-stage cancer proved no match for the rhetoric of activism.

As a result of the zeal that instituted PSA testing across the United States despite clear evidence of its harms and the conjectural nature of its benefits (and this even as the trend toward evidence-based medicine was gaining strength), tens of millions of American men have now been screened, with untold numbers undergoing treatment with its attendant harms, for a condition that would otherwise have remained latent. As if PSA testing had become a movement sustained by its own momentum, doubts voiced in 1989 over mass screening "for a cancer for which the mortality is far less than the histological incidence"[9] have been answered only with more screening. Precisely the same doubts were put into print in 2007.

In 2003, it was estimated that 1 in 6 men will be diagnosed with prostate cancer, and 1 in 34 will die from prostate cancer. Accordingly, given the low mortality rate, less than 1 in 5 men diagnosed with prostate cancer die of prostate cancer. Screening can tap into a very large reservoir of clinically silent prostate cancers. . . . [There exists] a tremendous gap between the presence of prostate cancer and death resulting from prostate cancer, and for every case of prostate cancer that would not have led to death or morbidity, any therapy administered is ineffective, unnecessary, and usually harmful.[10]

One measure of the reservoir's depth comes from the landmark Prostate Cancer Prevention Trial, which ran from 1993 to 2003, where an artificially intensive investigatory regimen discovered cancer in fully 24.4% of a placebo group originally classified as low-risk.[11] Clearly (and in striking analogy to the discovery of a vast and dubiously significant reservoir of depression), the more minutely you search for prostate cancer the more likely you are to find it, whether or not it possesses any actual medical significance.[12]

As we know, the placebo effect appears to be highly dependent on expectation. Perhaps there exists an investigative equivalent of placebo experience, whereby researchers believe they discover just what they expected and are buoyed accordingly. As it happens, the Prostate Cancer Prevention Trial which found an approximately 25% reduction in disease incidence in the treatment group over the course of the study was designed to detect a 25% difference in that variable with 92% power. The PCPT scored an expectation bull's-eye. Upon review, however, the FDA found the PCPT data considerably less impressive.[13] The principal investigator of the PCPT also stated in print a decade ago that the PSA revolution was progressing exactly according to the hopes and expectations of its movers (an enthusiasm he has since retracted).[14] The heady experience of having such predictions confirmed, or presuming them confirmed, communicates itself to patients in the form of confidence. Confident doctors in turn make for hopeful patients.

In part because of advocacy by confident doctors, and regardless of warnings against urological activism that have since proved well-founded, prostate-cancer screening was promoted so effectively that even the prospect of incontinence and impotence has not deterred men from seeking it out. The cancer detected with unprecedented frequency virtually as soon as PSA testing began seemed to confirm the urgent necessity for screening itself. Given that PSA enables the early discovery of prostate cancer, committed urologists assumed that mass screening could not but reduce mortality. With the means of early detection already in place, it only made sense, therefore, to use them. As for the risks of unleashing an epidemic of superfluous treatments with serious adverse effects, these, it seems, paled into insignificance next to the envisioned possibility of making prostate cancer a curable condition. They could be left to themselves. The constellation under which we now live—massive overdiagnosis and overtreatment of prostate cancer coupled with a kind of implicit conviction that PSA testing will prove itself, regardless of the shortage of confirming evidence[15]—was already taking shape by the mid-1990s.

Highly instrumental in the process was the use of public-relations methods to drive home the powerful but simplistic message that PSA saves lives: methods that had already proven themselves in the case of the sister disease, breast cancer. Not only, then, does the drive for the early detection of prostate cancer resemble concurrent efforts to identify depression in its early stages (when researchers went out into the community and found what they were looking for), but in another instance of social synergy, it was modeled on the drive to detect another cancer. Reflective of a strong mood of medical activism, all of these doublings produce an echo-chamber effect in which rhetoric achieves great resonance and people in large numbers find themselves reacting to it in the same way.

⬒

The PSA system as we know it could conceivably have been built from the ground up by urologists convinced that they did not have time to wait for the results of RCTs of PSA, given that deaths from prostate cancer stood at 40,000 per year in the U.S. when screening for the disease began. However, urologists did not invent every element of the system. Key components, including the rhetoric of early detection that drives men to get tested in the first place and vindicates the test no matter the result, were imported from breast-cancer medicine.[16] Historically speaking, mammography has led and PSA testing has followed, even though PSA lacks to this day the sort of validating evidence that mammography possessed before the PSA revolution began. One answer to the question "Why Is Prostate Cancer Screening So Common When the Evidence Is So Uncertain?"[17] is that those who launched PSA testing copied the successes of a screening mode for which evidence is less uncertain.

Isolated in 1979, PSA was at first used to monitor the progress of prostate cancer, not to identify the disease at an incipient stage. Prostate-cancer screening was introduced in the United States in 1987—when the use of mammography was already rising rapidly—to be followed in short order by a dramatic increase in detected cancer. By some estimates, from 1990 to 1991 alone the incidence of prostate cancer in the U.S. shot up 25%, and from 1990 to 1993, 60%. An iatrogenic epidemic was in the making. That numbers like these didn't deter men from getting screened but simply marked the beginning of the PSA revolution suggests a system lacking the braking effect of negative feedback.[18] Overdiagnosis as a direct result of mass screening leads to overtreatment, which leads to the appearance of saved lives, which in turn acts as a strong advertisement for screening: a "self-reinforcing" but socially powered cycle that "doesn't affect just the individuals tested but those who hear their stories as well, including friends, family members, and acquaintances."[19] And the drive to get men tested in the first place was patterned closely on the breast-cancer template of community-based screening and public "awareness"—in a word, the breast-cancer blueprint of a socio-medical movement. The promoters of PSA testing in the early 1990s did not start from zero but tapped into an existing model of proven efficacy. In order to ascertain why PSA testing is so common, we should bear in mind when and how it became common.

Somewhat in the tradition of the Utopians who are perfectly conscious of their own health and regard food and drink as "ways to withstand the insidious attacks of sickness,"[20] the screening movement puts us on guard against the insidious onset of cancer by raising consciousness. Between 1986 and 1989 (concurrent and resonant with the National Institute of Mental Health's Depression Awareness, Recognition, and Treatment Program), the American Cancer Society conducted a Breast Cancer Detection Awareness Program, the goal of which, according to a proponent, was "to make women and health professionals aware of the benefits of breast cancer detection."[21] Awareness, it seems, excluded knowledge of mammography's possible harms. A program to disseminate information about mammography's benefits and only the benefits (though the public receives a poor idea of their magnitude) is bound to produce a disconnection between enthusiasm for screening and the evidentiary record of the procedure itself. As it happens, the harms of such screening—not only false positives, but the detection of questionably significant lesions that are nonetheless treated with surgery or radiation—approximate those of PSA testing. Just as Breast Cancer Awareness supplied the precedent and template for Prostate Cancer Awareness, just as the selective understanding of awareness itself passed from the former to the latter, so the treatment of indolent forms of breast cancer under the banner of saving lives found a parallel in the PSA regime, which has somehow been reinforced, not discredited, by overdiagnosis and overtreatment.

In addition to an outpouring of mutually reinforcing articles in the medical literature, the Breast Cancer Detection Awareness Program generated an organized blizzard of pamphlets, television spots, news features, and newspaper inserts, all encouraging women to be screened. The principles of the program appear to have been to depict mass screening as a procedure without harms but with great benefits, to speak in the name of something unopposable (such as "education"), to reach people where they live, and to offer mammography at low cost. In Massachusetts, for example, a

> campaign entitled "Mammography: The Breast Test" was conducted to educate people about early detection of breast cancer. The program was held in late April when more than one million Massachusetts households received information on breast cancer during the residential crusade. The following week a toll-free number was available for information on low-cost mammograms ($50 or less) at more than 100 hospitals throughout the state.[22]

Well before Gen. Norman Schwarzkopf, the popular war hero, served as spokesman for Prostate Cancer Awareness, it seemed natural to use military metaphors like "campaign" and "crusade" in connection with breast cancer. The promotional campaign became the blueprint for Prostate Cancer Awareness. Indeed, Prostate Cancer Awareness Week began in the year the BCDAP ended—1989—and drew on the help of mammography advocates to get off the ground. "Several members of the [Prostate Cancer] Educational Council who had been associated with Breast Cancer Awareness Month contributed significantly" to the initiation of PCAW in 1989.[23] Defining itself as educational and delivering services in a community setting, PCAW was informed by the same principles as its predecessor. If human beings respond not only to evidence but to evocative associations, prostate-cancer awareness was associated with its breast-cancer counterpart by temporal proximity, rhetorical parallels, a similar orchestrated optimism, and analogous tactics of mobilization. And like Breast Cancer Awareness, PCAW caught on. By 1992, when the American Cancer Society endorsed PSA, free tests were being offered at 1800 clinics. Testing over three million men in the first decade of its existence, PCAW became the largest cancer screening program in the U.S.[24] It has since grown into Prostate Cancer Awareness Month.

Whereas the first trial of mammography—the Health Insurance Plan of Greater New York (HIP) trial, considered the first RCT in cancer screening—dates to 1963, randomized trials of PSA were initiated only in the 1990s. When PSA testing began there was therefore no body of evidence showing that it reduced mortality, which made it a sort of experiment on the male population, albeit without the constraint of informed consent that would operate in an actual experiment. But in getting around informed consent too PSA has followed mammography. Despite its professed educational mission, mammography relied on the use of public-relations methods to get the target population to the screening center. Success was measured by numbers screened, not by improvement of public understanding—a model that would govern Prostate Cancer Awareness as well. So it is that "the mammography controversy is a foreshadowing of . . . controversies about prostate cancer," as a critic of uninformed consent to mammography wrote in 1995 in the *Journal of the National Cancer Institute.*[25] The family resemblance between the uninformed woman and the uninformed man persists to this day. Where men getting PSA's tend to overestimate the risk of death from prostate cancer and presume that PSA reduces it, women often overestimate the risk of breast-cancer death and, while correctly assuming that mammography reduces it, greatly misjudge its effect.[26] In both cases many of the screened are unaware that screening also picks up what medicine knows as pseudo-disease—a term inconsistent with the lay understanding of cancer as either a lethal or potentially lethal, but certainly not an innocuous condition—and that such findings set off a cascade of consequences.

Although the evidence in favor of PSA testing falls short of that for mammography, the PSA regime was built on the mammography model and continues to resemble it. (Thus the paradox that while PSA has given rise to the harms of mammography without the demonstrated mortality benefit, it is defended in the language of risks and benefits.) But even the difference in supporting evidence becomes less salient in view of the common tendency among those screened to overrate benefits, often vastly, and to underestimate harms. So too, neither men nor women treated for screen-detected disease without clinical significance know this to be so (nor does medicine itself know in any given case, or else the patient wouldn't be treated); all they know is that their cancer has been treated, thus marking a win for the system that detected it. Because few would knowingly commit their body to a flawed system, such an arrangement presupposes some sacrifice of informed consent; but because the system could not keep going without willing entrants and public enthusiasm, the lack of informed consent must be masked as something honorific. The celebration of "awareness" meets this requirement. If you search Amazon for "cancer awareness," you will find knee socks, pens, lanyards, and stickers, but no books.

In 1997 the ACS qualified its recommendation of PSA testing for men over 50, now advising that candidates for the test be informed of its liabilities as well as benefits. Recognizing that the evidence for PSA testing was questionable, other bodies, including the American College of Physicians and the

American Academy of Family Physicians, called for informed consent around the same time. When the hopes and promises that inspired prostate-cancer "awareness" lost their luster and the evidentiary basis of PSA testing grew so questionable that one could speak with more assurance of harms than benefits, PSA testing became, in theory, the patient's decision. In practice it proceeded much as if no requirement for informed consent were in place.[27] Regardless of its evidentiary deficiencies, the test had already enshrined itself in American life; a sort of de facto presumption in its favor had set in. In 1999 the U.S. Postal Service issued a stamp with the messages, "Prostate Cancer Awareness" and "Annual Checkups and Tests," which suggests just how much of a movement PSA testing had already become. As I write this a decade later, a single morning brings the following headlines from Google News:

> Prostate Cancer Awareness Effort Continues
> NFL and Players Renew Commitment to Prostate Cancer Health
> Prostate Cancer Test Promoted
> Blue Cure Foundation Aims to Spread the Word About Prostate Cancer
>   Screening
> Atlanta Turns Blue
> Cancer Support Group Set to Spread Awareness
> Riverside Community Hospital Urges Prostate Cancer Testing
> Men Must Heed Prostate Cancer Risks
> Experts Recommend PSA Testing During Prostate Cancer Awareness Month

Despite the acknowledged harms of mass screening for prostate cancer, the movement thus continues unabated. Hence too the strong backlash against the recommendation against PSA testing by the U.S. Preventive Services Task Force in 2012. Only because of the reiterated message that early detection saves lives—as is true in the case of mammography, but may or may not be true in that of PSA—has the PSA system flourished despite the doubts that shadowed it from the beginning.

⊜

Like the antidepressant campaign, then, mass screening for prostate cancer draws power from publicity, in this case emanating from non-corporate sources and carrying an aura of civic activism. In both instances the publicity struck a nerve and set in motion a potent doubling effect whereby uptake of the advertised message acted as an advertisement in its own right. Both movements have been as much socially as medically driven, which means in the case of prostate-cancer screening that things like message-bearing post-age stamps and license plates,[28] contagious catchphrases, early-detection folklore, golf tournaments, public service announcements (those other PSA's), grabby headlines, blue ribbons, and the knowledge that friends, neighbors and co-workers get tested propel the movement at least as effectively as bet-ter sources of information. In both cases the publicity activating the move-ment misleads. The result is an artificial epidemic of depression on the one hand (the numbers on antidepressants having quadrupled since 1988) and prostate cancer on the other (now being detected with such frequency that the medical literature has begun to speak of a "risk of detection"). In both cases the epidemic, though exaggerated, had some basis in fact, which gave publicity something to bite on.

But these were not the only fin-de-siècle epidemics of dubious origin. "In the 1990s, the United States [became] the hot zone of psychogenic diseases, new and mutating forms of hysteria amplified by modern communications. . . . Infectious epidemics of hysteria spread by stories circulated through self-help books, articles in newspapers and magazines, TV talk shows and se-ries, films, the Internet, and even literary criticism."[29] Socially driven in large part and patterned on and reinforced by mammography, the PSA revolu-tion may also have acquired resonance from other anxious socio-medical movements competing for recognition at the same time, in the same place. Borne not by pathogens but by the mobilizing rhetoric of alarm, the epidem-ic of prostate cancer begins to resemble a psychogenic outbreak after all.[30]

Epilogue
Being Medically Unique

In a remarkable passage, Montaigne, the prince of skeptics, tells of a blind man who plays tennis:

> I saw a gentleman of a good family, born blind, or at least blind from an age such that he did not know what vision is; he understands so little what he lacks, that he uses and employs words proper to vision as we do and applies them in a way that is entirely his own and idiosyncratic. He was presented with a child to whom he was godfather. Taking it in his arms he said: "Oh, lord! What a lovely child! How beautiful it looks!" . . . There is more: since hunting, tennis, and shooting are our sports, and he has heard this said, he takes a liking to them, and busies himself with them, and believes he has the same part in them that we do. . . . He takes a tennis ball in his left hand and hits it with his racket; he shoots with his musket at random, and is satisfied when his people tell him he is too high or at the side.[1]

Though he "understands so little" that he appears to have convinced himself that he sees, it is not clear to me that the blind man should be dismissed as a fool. It could be said that finding himself in the Rome of the sighted, he simply does as the Romans do—takes part in human life as best he can.

Deriving the testimony of his senses from the reports of others ("How beautiful it looks!"), the sociable blind man resembles all who model their bodily experiences on those of others—who feel as others feel, according to the sociology of the placebo effect. We tend after all to experience what those like us do, or what we suppose they do, as when study subjects report less pain when it seems that others feel less,[2] or respond to the crafty prompt that the treatment they have been given has been shown to help others. The case has been made that because the information presented to a patient "potentially influences the experience of treatment outcomes,"[3] doctors need to be mindful of the placebo and nocebo effects that may spring from the very discussion of risks and benefits, which is another way of saying that reports of others' outcomes have an effect on our own.

While other factors, among them conditioning, may contribute to the placebo effect, it is largely as social animals that we enjoy its benefits, just as it is theorized to have evolved from social behavior such as mutual grooming among apes. By the same token, however, if we should lose the feeling that our case resembles others' and that we can be helped just as they have been, the entire edifice of hope and trust may break down, as I have learned from experience as a cancer patient. I don't mean to say the placebo effect has much to offer the cancer patient; the survival benefits for women with breast cancer who had "supportive-expressive group therapy" could not be replicated. I mean simply that my case has clarified for me the placebo effect's social sources. To be without the sense that one's case is like others is to play tennis at random, with no way of knowing whether a shot is in or out.

It all began conventionally enough when, like millions of others, I had a PSA test without the slightest idea of what I might be getting into. One test led to another and then to biopsies, one after another, my PSA rising all the while, until cancer was finally confirmed. Of the treatment options brachytherapy—the embedding of radioactive pellets or "seeds"—seemed the least bad, so that is what I chose. As it turned out, during the procedure the urologist was unable for some reason to get one string of seeds into place, but that didn't matter, we were told. Naturally I assumed the treatment was successful. But successful in doing what? As I became aware, belatedly, of the overdiagnosis of prostate cancer, I began to wonder if I had not unthinkingly sought treatment for an innocuous condition at the behest of medical activism.

But if mine was an innocuous condition it was also a stubborn one. Following the procedure my PSA dropped, but not enough, and as it resumed its upward course I became a patient in a medical center a thousand miles away, where a team of brisk, self-assured, not to say conceited doctors took over my case. "The patient who journeys to a famous clinic or physician is as ready to be helped as the pilgrim at a religious shrine,"[4] and in this case the first thing the priests did was form an idea about what was wrong with the pilgrim. Over time, and not without much theory-testing and diagnostic travail, I passed into the category of patients who have failed brachytherapy. Because radiation complicates surgical removal of the prostate if it should fail (as no one had explained to me), the best option at this point seemed to be a second procedure of the same kind, but more precisely targeted. My local doctor had left a "cold spot" that would now be correctly irradiated—this was the message. Five years after the original implant, a second was done.

In short order, however, the same sequence took place all over again—rising PSA, more biopsies with more samples (as many as 24 in one instance), imaging that yielded nothing, conferences, contradictions, delays, exploration and finally dismissal of the innocent possibility of PSA "bounce." Somewhere in the middle of this burlesque, the radiation oncologist threatened to cancel a brachytherapy unless I met with him beforehand—as it turned out, only so that he could demonstrate his mastery of my case to his residents. At another point, as I sat in the waiting room thumbing through a lifestyle magazine, I came across a feature about my own urologist's villa-like residence. In due course I became the only patient in his considerable experience, and I suspect one of the few to his knowledge, to have had not two but three brachytherapies. Three times I had to notify students that if any were pregnant they should not sit too close. With something like 150 spent radioactive pellets arrayed like chevrons in my prostate, meticulously placed but useless, I have gone from being one of countless men treated for prostate cancer—a rite of passage in the PSA era—to a data set of one.

A few years ago a paper concluded, "The best management of the small number of brachytherapy patients encountering failure is unclear at this time,"[5] referring of course to patients who fail once, not to the still smaller—the vanishingly small—number who fail twice. Now, with my PSA resuming its dismal pattern, the third treatment seems to have failed as well. A friend and former officer of the American Urological Association, himself a patient, but whose cancer is more advanced, warns that with three doses of radiation I am already "challenging morbidity" and advises against further biopsies. Another medical acquaintance, whose name is well known but whose uncommon kindness is not, says apologetically that at this point there is nothing he can do for me. My urologist, however, proposes resuming the same old round of biopsy and imaging, to be followed by a fourth brachytherapy if necessary. I am considering dutasteride, a drug whose incautious use to inhibit prostate cancer I have argued against.[6] In the midst of all this, I learned on good authority that a member of my team has recanted his activism, though when I asked him about it he refused to answer. I have since run across a paper of his on the benefits of group morale for the prostate cancer patient.

Over recent years as I have become an enigma to myself and others, my belief that medicine knows how to treat my case, or even understands it, has crumbled. I doubt my doctors know whether my cancer, invariably identified as Gleason 6 (Intermediate)—yet wrongly entered in the medical records at several points as Gleason 7—was ever significant. They are as blind as I am, though they would have me believe it is definitely significant because otherwise their treatments are harming me for nothing. Of one thing am I sure: their complete indifference not only to the contradiction in their or my records but to the side effects I have encountered, and which once landed me in the ICU, since I began this journey by following others only to end up being interesting to medicine in my own right. "Yours is a unique story," I have been told; but it is so involved, protracted and bewildering that it verges on the untellable, and in any event this is a matter in which no one wants to be unique.

Though as patients we hope to be treated compassionately, too much compassion can give us the wrong idea, and we may also prefer to be treated somewhat impersonally, if only because professionalism is reassuring and tells us our problem can be managed. Those who say doctors ought to show confidence are on to something. We don't want a doctor to act in a way that suggests he or she has never seen our case before. But what if he or she actually hasn't?

Now that it is clear to me that my urologist has in fact never seen my case or anything quite like it, hope and trust—those good companions—have been replaced by gnawing doubt. The man's professionalism, which might once have assured me that he knows exactly what he is doing, now seems a mask. His few words, which formerly made him seem less talker than doer, are now the shield of one who will not confide, admit or affirm anything. As my case has grown ever stranger and more intractable, his manner has remained exactly the same—frozen. He acts as if there were no reason to think a treatment that has failed repeatedly will not fail again, or that side effects are of any concern. It wouldn't bother him to learn, either, that many of the ill effects of repeated treatments, beginning with fatigue, cunningly mimic the markers of depression. The ritual meeting of patient and doctor has degenerated into a ceremony of repetition. There is no "therapeutic alliance." I am like someone in the placebo group of a study who is told, "This pill has been shown to help others," but understands this clever equivocation for what it is.

The doctors may have written off my case as inexplicable, but to me it has confirmed one thing at least: the largely social nature of the placebo effect. It is because of our bonds with others, including others whose experience seems like our own, that we are able to find sources of encouragement even in illness. Recall Haygarth's experiment at the Bath General Hospital in 1799, where subjects were specifically told that a certain worthless instrument had cured the pains of others and would cure theirs. It is easy enough to laugh at these credulous souls who wittingly or not modeled their very sensations on those of others, but we too are social beings, attuned to others. "The placebo reminds us that we are not alone."[7] Not to be able to liken one's experience to others is to be lost. As my case went from ordinary to incomprehensible and my sense that my experience resembled anyone else's melted away, I became lost indeed—blind, pathless.

But even as I went through treatment after treatment and trust in my doctors eroded and then collapsed, many showed the humanity they did not—nurses, clerical staff, intake and pre-op specialists, other doctors (some of great kindness), entire teams in the emergency room and the ICU. Like Telemachus in the palace of Menelaus, I came to them in need and found them not only attentive but generous in ways impossible to imagine before the event. One nurse made me forget my own humiliation as completely as if I had consumed an Egyptian drug. To her I was not a case but a member of the human company. To me she herself was heartsease.

Notes

Introduction

1   Henry Beecher, "The Powerful Placebo," *JAMA* 159 (1955): 1602-6.
2   Franklin Miller and Ted Kaptchuk, "The Power of Context: Reconceptualizing the Placebo Effect," *Journal of the Royal Society of Medicine* 101 (2008): 222-25.
3   Klaus Linde, Margrit Fässler and Karin Meissner, "Placebo Interventions, Placebo Effects and Clinical Practice," *Philosophical Transactions of the Royal Society B* 366 (2011): 1908.
4   Ned Stafford, "Germans Doctors are Told to Have an Open Attitude to Placebos," *BMJ* 342 (2011): 565.
5   Fabrizio Benedetti, "The Placebo Response: Science vs. Ethics and the Vulnerability of the Patient," *World Psychiatry* 11 (June 2012): 70. See also Franklin Miller, Luana Colloca, Ted Kaptchuk, "The Placebo Effect: Illness and Interpersonal Healing," *Perspectives in Biology and Medicine* 52 (2009), 518-539.
6   On the artificiality of laboratory studies of pain, see Henry Beecher, "Experimental Pharmacology and Measurement of the Placebo Response," *Science*, Aug. 15, 1952: 159; and Howard Spiro, *The Power of Hope: A Doctor's Perspective* (New Haven: Yale University Press, 1998), p. 50
7   Even so, "most of the studies on the placebo mechanism consider the mean reduction of a symptom in a group of subjects"; "the placebo effect is better described as a group effect." Fabrizio Benedetti and Luana Colloca, "Placebo-Induced Analgesia: Methodology, Neurobiology, Clinical Use, and Ethics," *Reviews in Analgesia* 7 (2004): 131.
8   Roger Ulrich, "View Through a Window May Influence Recovery from Surgery," *Science* 224 (1984): 420-21. The study took into account only those months when the trees were in leaf.
9   S. D. Goitein, *A Mediterranean Society: The Jewish Communities of the Arab World as Portrayed in the Documents of the Cairo Geniza* (Berkeley: University of California Press, 1967-1993), Vol. II, p. 241.
10 *The Praise of Folly and Other Writings*, ed. and tr. Robert M. Adams (New York: Norton, 1989), pp. 33-34.
11 Robert Burton, *The Anatomy of Melancholy* (New York: New York Review Books, 2001), I. 257. The foundational modern work of its kind, Thomas

Percival's *Medical Ethics* (1803), calls on physicians to "inspire the minds of their patients with . . . confidence," a provision incorporated into the original American Medical Association Code of Ethics and retained long thereafter. See Albert Jonsen, *The New Medicine and the Old Ethics* (Cambridge: Harvard University Press, 1990), p. 66.

12 Sherwin Nuland, *Maimonides* (New York: Schocken, 2005), p. 178.

13 No doubt the miscellaneous healers and wise women in the shadows of medicine also enjoyed reputation. The most celebrated occult healer of seventeenth-century England, one Valentine Greatrakes, "attracted hundreds of . . . sufferers upon whom he performed a number of successful cures." Keith Thomas, *Religion and the Decline of Magic* (New York: Scribner's, 1971), p. 203. On the placebo effect, see p. 209.

14 Frank Kermode, "'Opinion' in *Troilus and Cressida*," *Critical Quarterly* 54 (2012): 88-102.

15 P. Shaw, *The Reflector: Representing Human Affairs, As They Are; and May Be Improved* (London, 1750). Cited in Barry Blackwell et al., "Demonstration to Medical Students of Placebo Responses and Non-Drug Factors," *The Lancet*, June 10, 1972: 1280.

16 Throughout the era of pre-scientific medicine (and beyond) "the practitioner fitted the symptoms into a coherent, meaningful system, syntonic with the prevailing culture." Herbert Adler and Van Buren Hammett, "The Doctor-Patient Relationship Revisited: An Analysis of the Placebo Effect," *Annals of Internal Medicine* 78 (1973): 595-98.

17 Barron Lerner, *The Breast Cancer Wars: Hope, Fear, and the Pursuit of a Cure in Twentieth-Century America* (Oxford: Oxford University Press, 2001), pp. 272, 274.

18 Irving Kirsch, *The Emperor's New Drugs: Exploding the Antidepressant Myth* (New York: Basic, 2010), p. 165.

19 Allan Horwitz and Jerome Wakefield, *The Loss of Sadness: How Psychiatry Transformed Normal Sorrow into Depressive Disorder* (New York: Oxford University Press, 2007), e.g., pp. 26, 138.

20 Report issued by the National Center for Health Statistics, Oct. 19, 2011.

21 Michel de Montaigne, "On the Power of the Imagination," *Complete Essays*, tr. M. A. Screech (London: Penguin, 2003), p. 109.

22 Cited in Harold Merskey, *The Analysis of Hysteria, Second Edition: Understanding Conversion and Dissociation* (London: Gaskell, 1995), p. 8. Cf. Burton's observation that "men, if they but see another man tremble, giddy, or sick of some fearful disease, their apprehension and fear is so strong in this kind

that they will have the same disease." *Anatomy of Melancholy* I.255.

23   William Falconer, *A Dissertation on the Influence of the Passions Upon Disorders of the Body* (London, 1788), p. 72: "The hysteric paroxysm . . . is extremely apt to recur on the sight of people so affected. I once had an opportunity of seeing an instance of this kind at one of the publick water-drinking places in this kingdom. A lady was seized with hysteric convulsions during the time of divine service. In less than a minute, six persons were affected in a similar manner."

24   Lene Vase et al., "Patients Direct Experiences as Central Elements of Placebo Analgesia," *Philosophical Transactions of the Royal Society: Biological Sciences* 366 (2011): 1913-21.

25   M. M. Bakhtin, *The Dialogic Imagination,* tr. Caryl Emerson and Michael Holquist (Austin: University of Texas Press, 1981), p. 338.

26   On this possibility see Gunver Kienle and Helmut Kiene, "The Powerful Placebo Effect: Fact or Fiction?", *Journal of Clinical Epidemiology* 50 (1997): 1311-18. A criticism sometimes made of open, that is, non-blind placebo studies is that subjects may report benefits in order to give the experimenters what they are looking for.

27   Bruce Barrett et al., "Placebo, Meaning, and Health," *Perspectives in Biology and Medicine* 49 (2006): 189.

28   Roy Porter, *English Society in the 18th Century* (London: Penguin, 1991), pp. 226-27.

29   Possibly the first treatise on coffee was written by the same man whose translation of the *Thousand and One Nights* into French a few years later marks a transforming event in literary history. See Marina Warner, *Stranger Magic: Charmed States and the Arabian Nights* (Cambridge, MA: Harvard University Press, 2012), p. 13.

30   A. Branthwaite and P. Cooper, "Analgesic Effects of Branding in Treatment of Headaches," *British Medical Journal* 282 (1981): 1576-78.

31   Irving Kirsch, "Response Expectancy as a Determinant of Experience and Behavior," *American Psychologist* 40 (1985): 1191.

32   Barrett et al., "Placebo, Meaning, and Health,": 184. Cf. Daniel Moerman and Wayne Jonas, "Deconstructing the Placebo Effect and Finding the Meaning Response," *Annals of Internal Medicine* 136 (2002): 471-76; and Arif Khan et al., "Are the Colors and Shapes of Current Psychotropics Designed to Maximize the Placebo Response?", *Psychopharmacology* 211 (2010): 113-22.

33   Blackwell et al., "Demonstration to Medical Students of Placebo Responses and Non-Drug Factors": 1279-82. On the social inflection of moods, see

Vincent Nowlis and Helen Nowlis, "The Description and Analysis of Mood," *Annals of the New York Academy of Sciences* 65 (1956): 345-55. According to this study, subjects' reports of their own mood vary with the drugs taken by those around them, so that someone taking dramamine amidst others taking seconal reports different mood indicators from someone taking dramamine amidst similar others.

34 Barbara Duden, *The Woman Beneath the Skin: A Doctor's Patients in Eighteenth-Century Germany*, tr. Thomas Dunlap (Cambridge: Harvard University Press, 1991), pp. 90, 130.

35 Cited in Gilbert Honigfeld, "Non-Specific Factors in Treatment," *Diseases of the Nervous System* 25 (1964): 150.

36 Allan Horwitz, *Creating Mental Illness* (Chicago: University of Chicago Press, 2002), p. 109.

37 Ted Kaptchuk, "Powerful Placebo: The Dark Side of the Randomised Controlled Trial," *Lancet* 351 (June 6, 1998): 1722-25.

38 Howard Spiro, *Doctors, Patients, and Placebos* (New Haven: Yale University Press, 1986), p. 1.

39 J. L. Mommaerts and Dirk Devroey, "The Placebo Effect: How the Subconscious Fits In," *Perspectives in Biology and Medicine* 55 (2012): 53; Howard Spiro, *The Power of Hope: A Doctor's Perspective* (New Haven: Yale University Press, 1998), p. 52.

40 Ralph Horwitz and Sarah Horwitz, "Adherence to Treatment and Health Outcomes," *Archives of Internal Medicine* 153 (1993): 1863: "In numerous studies, patients who adhere to treatment, even when that treatment is a placebo, have better health outcomes than poorly adherent patients." Cf. Ira Wilson, "Adherence, Placebo Effects, and Mortality," *Journal of General Internal Medicine* 25 (2010): 1270-72; and Andrew Avins et al., "Placebo Adherence and Its Association with Morbidity and Mortality in the Studies of Left Ventricular Dysfunction," *Journal of General Internal Medicine* 25 (2010): 1275-81.

41 Adler and Hammett, "The Doctor-Patient Relationship Revisited": 596.

42 Spiro, *Power of Hope*, p. 251. Lenin surrendered his passion of smoking in order to devote himself to revolution.

43 Jonathan Shay, *Odysseus in America: Combat Trauma and the Trials of Homecoming* (New York: Scribner, 2002), p. 5.

44 Carol Kronenwetter et al., "A Qualitative Analysis of Interviews of Men with Early-Stage Prostate Cancer: The Prostate Cancer Lifestyle Trial," *Cancer Nursing* 28 (2005): 99-107.

45 If people appear to stop smoking in groups, it is also true that the tobacco

industry originally "advertised cigarettes as a form of social glue that would 'stick' individuals into cohesive groups." Siddhartha Mukherjee, *The Emperor of All Maladies: A Biography of Cancer* (London: Fourth Estate, 2011), p. 445. Social networks support both the termination and beginning of smoking.

46   Spiro, *Doctors, Patients, and Placebos*, e.g., pp. 220-22.

47   An analysis that may last years gives ample scope for variables other than the analysis itself to operate.

48   Arthur Shapiro and Elaine Shapiro, *The Powerful Placebo: From Ancient Priest to Modern Physician* (Baltimore: Johns Hopkins University Press, 1997), p. 53.

49   Shapiro and Shapiro, *The Powerful Placebo*, p. 2.

50   *Gargantua and Pantagruel*, tr. M. A. Screech (London: Penguin, 2006), pp. 640-41.

51   *Gargantua and Pantagruel*, p. 640.

52   *The Decameron*, tr. G. H. McWilliam (London: Penguin, 1995), p. 660. In Chaucer's Merchant's Tale, the wily May convinces her husband that he recovered his sight as a result of her "medicine." The tale features a figure named Placebo who ministers to his brother's (the husband's) fantasies. In the *Arabian Nights* tale of "Aladdin of the Beautiful Moles," "one of the richest and most powerful merchants in Baghdad grieves that he has no children, and a passing magician rallies him by giving him a wonderful mixture of stuff to thicken his sperm. The potion is a placebo—its power lies wholly in the believing, and the merchant does believe. A baby is born." Warner, *Stranger Magic*, p. 403.

53   James House et al., "Social Relationships and Health," *Science* 241 (1988): 540-45. For a popular version see Dean Ornish, "Love Is Real Medicine," *Newsweek*, Oct. 3, 2005. On social bonds and health see also William Ruberman et al., "Psychosocial Influences on Mortality After Myocardial Infarction," *New England Journal of Medicine* 311 (1984): 552-59; and Farouk Mookadam and Heather Arthur, "Social Support and Its Relation to Morbidity and Mortality After Acute Myocardial Infarction," *Archives of Internal Medicine* 164 (2004): 1514-18.

54   Horwitz and Wakefield, *Loss of Sadness*, pp. 10-11. At one point shortly before his terrible illness, Tolstoy's Ivan Ilych is visited with what we are told is intolerable depression. There is a story behind it—the story of his life.

55   Wallace Stegner, *The Spectator Bird* (New York: Penguin, 1976), p. 109.

Chapter One

1   Homer, *The Odyssey*, tr. Richmond Lattimore (New York: HarperCollins, 1991), 4: 220-32.

2   Vivian Nutton, *Ancient Medicine* (London: Routledge, 2004), p. 41.

3   However, Circe herself attributes Odysseus's resistance to her drugs to the power of his mind: "There is a mind in you no magic will work on" (10.329).

4   Jasper Griffin, *Homer on Life and Death* (Oxford: Oxford University Press, 1983), p. 191. Cf. Odyssey 4.805. Menelaus himself will end not in Hades but in the Elysian Field, site of "the easiest life for mortals" (4.565).

5   Note, though, that it is not with this evil compound but with her wand that Circe transforms the men into swine.

6   That a story cannot be told properly without the observance of social rules is memorably illustrated by work of fiction composed by a physician a century ago: Chekhov's "Misery."

7   Hannah Arendt, *The Human Condition* (Chicago: University of Chicago Press, 1957), p. 175.

8   As if Zeus himself bowed to the dictates of storytelling, he brings the Odyssey to an end in Book 24 by causing the incensed kinsmen of the slaughtered suitors to forget their death. Unless such a resolution, as improbable as it is, had been imposed from above, the massacre of the suitors would have been the beginning of Odysseus's troubles and not the end of them. Zeus announces his decision to Athena by saying, "Let us make them forget the death of their brothers / and sons, and let them be friends with each other, as in times past" (24.484-85). It is as if the suitors' kinsmen were henceforth to be under the influence of heartsease, said to make a man oblivious to the death of a loved one; or more plausibly (for the ending of the Odyssey is hard to credit), the influence of heartsease resembles the magic of storytelling itself, the power responsible for the shape of the Odyssey as a completed work.

9   Plato, *Laws*, tr. T. Saunders (London: Penguin, 1975), p. 312.

10   Between a benign drug and a poison, only the former, it seems, has a social component. When Telemachus prepares to sail for Pylos in search of news of his father, one of the suitors speculates sarcastically that he may bring back "poisonous medicines / and put them into our wine bowl, and so destroy all of us" (2.329-30). Unlike the drug introduced by Helen into the wine bowl, this speculative poison will be administered secretly and its action will not be supported by ritual.

11   When the boy Odysseus is gored by a boar, the wound is bound up with "in-

cantations" (19.457).

12   "The whole [Homeric] epic is in a way an enthusiastic homage to superiority in the use of words and their power to touch men's hearts." Pedro Lain Entralgo, *The Therapy of the Word in Classical Antiquity*, tr. L. J. Rather and John Sharp (New Haven: Yale University Press, 1970), p. 29.

13   Howard Spiro, *The Power of Hope: A Doctor's Perspective* (New Haven: Yale University Press,1998), p. 228.

14   Plato, *Charmides*, tr. Thomas West and Grace Starry West (Indianapolis: Hackett, 1986), p. 18.

15   Spiro, *The Power of Hope*, pp. 210, 226.

16   Fabrizio Benedetti, *Placebo Effects: Understanding the Mechanisms in Health and Disease* (Oxford: Oxford University Press, 2009), p. 39.

17   Cf. Hannah Arendt, *Totalitarianism*; Part Three of *The Origins of Totalitarianism* (New York: Harcourt Brace Jovanovich, 1951), p. 174: "For the confirmation of my identity I depend entirely upon other people; and it is the great saving grace of companionship for solitary men that it makes them 'whole' again, . . . restores the identity which makes them speak with the single voice of one unexchangeable person." Note the language of healing.

18   Connie Peck and Grahame Coleman, "Implications of Placebo Theory for Clinical Research and Practice in Pain Management," *Theoretical Medicine* 12 (1991): 265.

19   Franklin Miller and Ted Kaptchuk, "The Power of Context: Reconceptualizing the Placebo Effect," *Journal of the Royal Society of Medicine* 101 (2008): 225. See also the more ironic formulation of Keith Thomas, *Religion and the Decline of Magic* (New York: Scribner's, 1971), p. 667: "Sociologists have observed that contemporary doctors and surgeons engage in many ritual practices of a non-operative kind. Modern medicine shares an optimistic bias with the charmers and wise women and it has similar means of explaining away failure."

20   "Perhaps only when a friend, relative, or healer indicates some level of social support (for example, by performing a ritual) is the individual's internal [healing] economy able to act." Daniel Moerman and Wayne Jonas, "Deconstructing the Placebo Effect and Finding the Meaning Response," *Annals of Internal Medicine* 136 (2002): 475.

Chapter Two

1   James House et al., "Social Relationships and Health," *Science* 241 (1988): 544.

2   More, *Utopia*, tr. George Logan and Robert M. Adams (Cambridge: Cambridge University Press, 1989), p. 61.

3   Plato, *Laws*, tr. T. Saunders (London: Penguin, 1975), p. 449.

4   *Utopia* is framed by the language of medicine. Measures regulating wealth without abolishing it "may have as much effect as good and careful nursing has on persons who are chronically sick. The social evils I mentioned may be alleviated and their effects mitigated for a while, but so long as private property remains, there is no hope at all of effecting a cure and restoring society to good health" (p. 39). Private property is abolished in Utopia.

5   Cf. William Morris's Nowhere, whose inhabitants, aglow with health, boast of their collective wellbeing.

6   *Laws*, p. 347.

7   A. Alvarez, *The Savage God: A Study of Suicide* (London: Norton, 1990), p. 79.

Chapter Three

1   Robert Burton *The Anatomy of Melancholy*. Three vols. in one. (New York: New York Review Press, 2001), II. 119. On "that true nepenthe of Homer, which was no Indian plant," see II.112. Cf. the mock inventory of the utopian fields in Erasmus, *The Praise of Folly and Other Writings* tr. Robert M. Adams (New York: Norton, 1989), pp. 10-11: Folly was "born on the Fortunate Isles, where all things grow 'unsown and uncultivated.' In that part of the world nobody works, grows old, suffers from sickness; the fields bear no day-lilies, mallow, leeks, beans, or vulgar vegetables of that sort. But everywhere eyes and noses are gratified with moly, heal-all, nepenthe, marjoram, ambrosia, lotus, roses, violets and hyacinths, as in the garden of Adonis."

2   Though the book does begin by telling of a wedding feast, complete with dancers and acrobats.

3   Marie Prévost, Anna Zuckerman, and Ian Gold, "Trust in Placebos," *Journal of Mind-Body Regulation* 1 (2011): 141.

4   Rabelais, *Gargantua and Pantagruel*, tr. M. A. Screech (London: Penguin, 2006), p. 639.

5   Burton, *The Anatomy of Melancholy*, II.119.

6   This sort of advice is not as trivial as it may seem. Recently it was pointed out that physical exercise alleviates depression—without the side-effects of drugs. Irving Kirsch, *The Emperor's New Drugs: Exploding the Antidepressant Myth* (New York: Basic, 2010), pp. 169-73. Inactivity, one of the telltale symptoms of depression, also gives Hamlet his most famous trait—procrastination.

7   On traditional attitudes toward depression (melancholy), see Allan Horwitz, *Creating Mental Illness* (Chicago: University of Chicago Press, 2002), pp. 28-29.

8   On the stereotypical nature of the King's recommendations, see Bridget Gellert Lyons, *Voices of Melancholy: Studies of Literary Treatments of Melancholy in Renaissance England* (New York: Norton, 1975), pp. 86-87.

9   *Anatomy of Melancholy*, III. 432. Cf. Judith Kegan Gardiner, "Elizabethan Psychology and Burton's *Anatomy of Melancholy*," *Journal of the History of Ideas* 38 (1977): 380.

10   All references are to the Norton Shakespeare, ed. Stephen Greenblatt, Walter Cohen, Jean E. Howard, and Katharine Eisaman Maus (New York: Norton, 2007).

11   Frank Kermode, *Shakespeare's Language* (New York: Farrar, Straus and Giroux, 2000), p. 103.

12   Avoidance of excessive study is also prescribed in Bright's *Treatise of Melancholy* (1586), a standard text at the time.

13   Eric Langley, "Plagued by Kindness: Contagious Sympathy in Shakespearean Drama," published online in *Medical Humanities*, Sept. 2, 2011 in advance of print; doi:10.1136/medhum-2011-010039. See p. 3. Iago enacts "a cruel parody of the medicinal friendship": p. 5.

14   Mikhail Bakhtin, *Problems of Dostoevsky's Poetics*, tr. Caryl Emerson (Minneapolis: University of Minnesota Press, 1984), p. 59.

15   Roger Scruton, *Beauty* (New York: Oxford University Press, 2009), p. 94.

Chapter Four

1   Hume, *A Treatise of Human Nature* (Harmondsworth, Middlesex: Penguin, 1969), p. 169.

2   John Leonard, *Naming in Paradise: Milton and the Language of Adam and Eve* (Oxford: Clarendon Press, 1990), p. 199: "The serpent's single most persuasive argument is his ability to argue."

3   Fabrizio Benedetti, *Placebo Effects: Understanding the Mechanisms in Health and Disease* (Oxford: Oxford University Press, 2009), e.g., p. 84.

4   Cf. Bacon on the dynamics of suggestion: "As in Infection, and Contagion from Body to Body, (as the Plague, and the like) . . . the Infection is received by the Body Passive . . . so much more in Impressions from Minde to Minde, or from Spirit to Spirit, the Impression [will] be tak[en by] the Minde and Spirit, which is Passive . . . and therefore, they worke upon Weake Mindes . . . as those of Women. . . . The Cause of this Successe is to be truly ascribed unto the Force of Affection and Imagination, upon the Body Agent." Cited in Eric Langley,

"Plagued by Kindness: Contagious Sympathy in Shakespearean Drama," published online in *Medical Humanities*, Sept. 2, 2011 in advance of print; doi:10.1136/medhum-2011-010039. See p. 4.

5   The commentator is one Thyer ("the Librarian at Manchester") cited in Thomas Newton's 1763 edition of *Paradise Lost* at 9.794.

6   Cf. Northrop Frye, *The Return of Eden: Five Essays on Milton's Epics* (Toronto: University of Toronto Press, 1965), p. 78: Eve "repeats Satan's arguments as though they were her own."

7   See Barbara Lewalski, "Milton and Idolatry," *SEL* 43 (2003): 213-32.

8   "Areopagitica" in Milton, *Complete English Poems, Of Education, Areopagitica* (London: Everyman, 1990), pp. 604, 605, 615.

9   John Haygarth, *Of the Imagination as a Cause and as a Cure of Disorders of the Body* (Bath: Cruttwell, 1800), p. 29. In Tolstoy's great medical fable *The Death of Ivan Ilych* the protagonist derives some momentary comfort from a treatment or diagnosis, even from wonder-working icons and the last rites, only to see the sensation of relief evaporate.

10   Haygarth, *Of the Imagination as a Cause*, p. 12.

11   Charles Taylor, *Sources of the Self: The Making of the Modern Identity* (Cambridge, Mass.: Harvard University Press, 1989), e.g., p. 227.

12   Taylor, *Sources of the Self*, p. 216.

13   Benjamin Franklin, *Writings* (New York: Library of America, 1987), pp. 1198, 1202. Many of Poor Richard's sayings affirm marriage.

Chapter Five

1   Rambler No. 4.

2   Vivian Nutton, *Ancient Medicine* (London: Routledge, 2004), p. 102.

3   Nutton, *Ancient Medicine*, pp. 239, 242, 121.

4   The charlatanism of public healing attaches to Verena Tarrant in Henry James' *The Bostonians* (1886). "'The daughter of Doctor Tarrant, the mesmeric healer—Miss Verena. She's a high-class speaker.' 'What do you mean?' Olive asked. 'Does she give public addresses?' 'Oh yes, she had quite a career in the West. I heard her last spring in Topeka. They call it inspirational. I don't know what it is—only it's exquisite, so fresh and poetical. She has to have her father to start her up. It seems to pass into her.'" *The Bostonians* (New York: Knopf, 1992), p. 44.

5   Robert Darnton, *Mesmerism and the End of the Enlightenment in France*

(Cambridge, Mass: Harvard University Press, 1968), p. 40.

6   Darnton, *Mesmerism*, p. 117.

7   Jessica Riskin, *Science in the Age of Sensibility: the Sentimental Empiricists of the French Enlightenment* (Chicago: University of Chicago Press, 2002), p. 192.

8   In the seventeenth century certain healers sought to cure by "stroking" in such a way as to draw the cause of illness out of the body. "Stroking could be represented as a magnetic means of easing the evil humours down through the limbs and out through the extremities." Keith Thomas, *Religion and the Decline of Magic* (New York: Scribner's, 1971), p. 204. In accordance with the fancies of an age of reason, Perkins heals not by laying on hands, not by virtue of personal charisma or a gift from God, but by a kind of impersonal stroking.

9   Benjamin Douglas Perkins, *The Influence of Metallic Tractors on the Human Body* (London: J. Johnson, 1798), pp. 47, 50, 53.

10   James Delbourgo, "Common Sense, Useful Knowledge, and Matters of Fact in the Late Enlightenment: the Transatlantic Career of Perkins's Tractor," *William and Mary Quarterly*, 3rd Series, 61 (2004): 643-684.

11   Perkins, *The Influence of Metallic Tractors on the Human Body*, pp. 83-84.

12   Claude-Anne Lopez, "Franklin and Mesmer: an Exchange," *Yale Journal of Biology and Medicine* 66 (1993): 329.

13   Benjamin Franklin, *Writings* (New York: Library of America, 1987), p. 788.

14   John Haygarth, *Of the Imagination as a Cause and as a Cure of Disorders of the Body* (Bath: Cruttwell, 1800), p. 4.

15   Haygarth, *Of the Imagination as a Cause and as a Cure of Disorders of the Body*, p. 4.

16   Haygarth, *Of the Imagination as a Cause and as a Cure of Disorders of the Body*, p. 9.

17   Haygarth, *Of the Imagination as a Cause and as a Cure of Disorders of the Body*, p. 7.

18   Rambler No. 4.

19   Mike Jay, *The Atmosphere of Heaven: The Unnatural Experiments of Dr Beddoes and His Sons of Genius* (New Haven: Yale University Press, 2009), pp. 177, 178, 213. A colleague of Haygarth's in Bristol reports that Beddoes lent him a set of Perkins tractors.

20   Irving Kirsch, *The Emperor's New Drugs: Exploding the Antidepressant Myth* (New York: Basic, 2010), p. 126.

21   S. Karen Chung et al., "Revelation of a Personal Placebo Response: Its Effects on Mood, Attitudes and Future Placebo Responding," *Pain* 132 (2007): 281-88.

22   Ted Kaptchuk et al., "Placebos Without Deception: a Randomized Controlled

Trial in Irritable Bowel Syndrome," *PLoS ONE* 5(12): e 15591.

23   A. Branthwaite and P. Cooper, "Analgesic Effects of Branding in Treatment of Headache," *British Medical Journal* 282 (1981): 1576-78.

24   N. McKendrick, "Josiah Wedgwood: An Eighteenth-Century Entrepreneur in Salesmanship and Marketing Techniques," *Economic History Review*, New Series 12 (1960): 408-33.

25   A version of this chapter appeared in *Medical Humanities* 37 (2011): 34-37. Published online in advance of print.

Chapter Six

1   On the Perkins tractor see Ulrich Tröhler, *"To Improve the Evidence of Medicine": The 18th Century British Origins of a Critical Approach* (Edinburgh: Royal College of Physicians, 2000); and David Wootton, *Bad Medicine: Doctors Doing Harm Since Hippocrates* (Oxford: Oxford University Press, 2006), pp. 166-170.

2   Francis Lobo, "John Haygarth, Smallpox and Religious Dissent in Eighteenth-Century England" in *The Medical Enlightenment of the Eighteenth Century*, eds. Andrew Cunningham and Roger French (Cambridge: Cambridge University Press, 1990), pp. 217-53.

3   Ordinarily Haygarth would have treated rheumatism with cinchona (quinine), now recognized as the first specific drug, though specific against malaria, not rheumatism. A few years after his exposé of the Perkins tractor Haygarth went on to publish *A Clinical History of Acute Rheumatism*.

4   John Haygarth, *Of the Imagination as a Cause and as a Cure of Disorders of the Body; as Exemplified by Fictitious Tractors and Epidemical Convulsions* (Bath: Cruttwell 1800), pp. 3-4. On the reception of Haygarth's argument see Christopher Booth, *John Haygarth, FRS (1740-1827): A Physician of the Enlightenment* (Philadelphia: American Philosophical Society, 2005), pp. 106-07.

5   In the London Morning Herald, December 25, 1799, one Dr. John Mather, Member of the Royal College of Physicians, London, wrote that "He cannot but think it his duty to adopt any remedy which attentive observation and experience assure him is eminently calculated to relieve the afflicted. In a variety of diseases, especially those of a topical kind, he therefore purposes to add to the usual remedies, Dr. PERKINS' METALLIC TRACTORS, which probably may most successfully be applied by a Medical hand. Though they do no harm even where they do no good, he has sufficient reason to believe that they

possess great powers, when a proper discrimination is made as to the nature of the case." The wording of the last sentence is curious and possibly erroneous, but suggests that one of the attractions of the Perkins treatment was its sheer innocuousness. Before "do no harm" became a guiding principle or at least motto of modern medicine, it was a selling point for a sham treatment. Recommending a certain concoction for rheumatism, Dr. Johnson cannot say if it will work—he knows of its use only in one case—but thinks it worth trying as long as it does no harm. See James Boswell, *The Life of Samuel Johnson* (New York: Knopf, 1992), p. 551.

6   Arthur K. Shapiro and Elaine Shapiro, *The Powerful Placebo: From Ancient Priest to Modern Physician* (Baltimore: Johns Hopkins University Press, 1997), p. 25. Haygarth himself bled, one case being recorded in Booth, John Haygarth, p. 41.

7   C. E. Kerr, I. Milne, and T. Kaptchuk, "William Cullen and a Missing Mind-Body Link in the Early History of Placebos," *Journal of the Royal Society of Medicine* 101 (2008): 89-92.

8   In an experiment similar to his own, one of Haygarth's correspondents plays "the part of a necromancer," tracing geometric figures with false tractors made of ten-penny nails. Haygarth, *Of the Imagination*, p. 17.

9   Robert Burton, *The Anatomy of Melancholy* (New York: New York Review Books, 2001), I. 256-57. Putting to one side the matter of faith-healing, Hamlet suggests at many points that imagination can act on the body. Simply by telling his tale the Ghost can make hair stand up, and Hamlet stages a simulation of the murder of his father to see if it will make the King flinch. He more than flinches.

10   The belief that a murder victim lives on in some way to testify against the killer was almost certainly known to Shakespeare. See Malcolm Gaskill, "Reporting Murder: Fiction in the Archives of Early Modern England," *Social History* 23 (1998): 1-30. In *Richard III* Lady Anne exclaims that "dead Henry's [that is, Henry VI's] wounds / Open their congeal'd mouths and bleed afresh" (1.2.55-56) in the presence of his murderer, Richard. However, the blood referred to is a flight of hyperbole—a poetic index of the extremity of Richard's evil—not the manifestation of an outraged universe. For the interesting case of a loaf of bread that bleeds to indicate a murderer, see Natalie Zemon Davis, *Fiction in the Archives: Pardon Tales and Their Tellers in Sixteenth-Century France* (Stanford: Stanford University Press, 1987), p. 66.

11   In Darwin the apprehended woman, "like a witch in a play," calls on Heaven to grant "that thou never mayest know again the blessing to be warm." Eras-

mus Darwin, *Zoonomia; or, The Laws of Organic Life* (New York: AMP Press, 1974; orig. pub. 1796), II.359. The farmer "mistakes . . . imaginations for realities" (II.356).

12   This in spite of the rhetoric of transparency and verifiable fact in which the Perkins tractor was advertised. On Haygarth's demand for evidence, see his letter (quoted in Lobo, "John Haygarth, Smallpox and Religious Dissent in Eighteenth-Century England": 7) to an associate of Jenner cautioning that "very full and clear evidence will be required" if Jenner's discovery of "vaccine inoculation" is to be believed. In time Haygarth was won over.

13   Haygarth, *Of the Imagination*, p. 12.

14   Haygarth, *Of the Imagination*, p. 1. Mark Akenside was a physician and author of a long poem on The Pleasures of Imagination, written at the age of twenty-three and inspired by Addison. The phrase "The Pleasures of the Imagination" appears in Spectator No. 411.

15   Spectator 411.

16   Himself afflicted with melancholy, Johnson deemed the *Anatomy of Melancholy* a work of "great spirit and great power." Boswell, *Life of Samuel Johnson*, p. 607.

17   A meaning not recorded in Johnson's dictionary but current in his time and indeed used by himself in Rambler No. 43, as cited in the *OED*. The first definition of Imagination given by Johnson is "Fancy . . . the power of representing things absent to one's self or others."

18   The astronomer of Rasselas is also cited in Lorraine Daston, "Fear and Loathing of the Imagination in Science," *Daedalus* 127 (1998): 77.

19   Boswell, *Life of Samuel Johnson*, p. 867. Even the well-known definition of imagination in Rambler No. 60 casts that faculty as an agency of deception: "All joy or sorrow for the happiness or calamities of others is produced by an act of the imagination, that realizes the event, however fictitious, or approximates it, however remote, by placing us, for a time, in the condition of him whose fortune we contemplate; so that we feel, while the deception lasts, whatever motions would be excited by the same good or evil happening to ourselves."

20   See Lobo, "John Haygarth, Smallpox and Religious Dissent in Eighteenth-Century England": 220.

21   Adam Smith, *The Theory of Moral Sentiments* (Indianapolis: Liberty Classics, 1982), pp. 50-51.

22   Smith, *Theory of Moral Sentiments*, pp. 51-52.

23   Smith, *Theory of Moral Sentiments*, p. 183.

24 Smith, *Theory of Moral Sentiments*, p. 9. On sympathy, mind and body, cf. Kerr, Milne, and Kaptchuk, "William Cullen and a Missing Mind-Body Link in the Early History of Placebos."

25 William Falconer, *A Dissertation on the Influence of the Passions upon Disorders of the Body* (London: Dilly and Phillips, 1788), p. 48. On the complementary case of a condemned man who died by force of imagination on the scaffold, thereby sparing himself, see Montaigne, "On the Power of the Imagination," in *Complete Essays*, ed. M. A. Screech (London: Penguin, 2003), p. 110.

26 Falconer, *A Dissertation on the Influence of the Passions upon Disorders of the Body*, p. 23.

27 When Haygarth proposed his experiment to Falconer, the latter "entirely approved the idea, and very readily consented to make the proposed trial upon the most proper cases which could be selected from his patients in the General Hospital" (*Of the Imagination*, p. 2).

28 Falconer, *A Dissertation on the Influence of the Passions upon Disorders of the Body*, pp. 23, 51.

29 Carl Woodring, ed., *Prose of the Romantic Period* (Boston: Houghton Mifflin, 1961), p. 65.

30 Clearly this was a mass phenomenon, as if a revival of morale were passing from person to person, reinforcing itself as it went; or as if a multiplier effect set in as more and more experienced what others seemed to. Suddenly recovery itself had become contagious.

31 Haygarth, *Of the Imagination*, p. 30.

32 E.g., A. Sandler, C. Glesne and G. Geller, "Children's and Parents' Perspectives on Open-Label Use of Placebos in the Treatment of ADHD," *Child: Care, Health and Development* 34 (2008): 118; A. Campbell, "Hidden Assumptions and the Placebo Effect," *Acupuncture in Medicine* 27 (2009): 68-69.

33 René Descartes, *Discourse on Method*, tr. Laurence Lafleur (Upper Saddle River, NJ: Library of Liberal Arts, 1956), p. 4.

34 L. J. Rather cited in Harold Merskey, *The Analysis of Hysteria, Second Edition: Understanding Conversion and Dissociation* (London: Gaskell, 1995), p. 11.

35 See Donald Bruce, *Radical Dr. Smollett* (Boston: Houghton Mifflin, 1965), ch. 4. Cf. Montaigne, "On the Power of the Imagination," p. 117: "I know of a squire who had entertained a good company in his hall and then, four or five days later, boasted as a joke (for there was no truth in it) that he had made them eat cat pie; one of the young ladies in the party was struck with such hor-

ror at this that she collapsed with a serious stomach disorder and a fever: it was impossible to save her." On "the close stitching of mind to body," see p. 118.

36   *Humphry Clinker* (London: Penguin, 1985), p. 33. Cf. p. 187: "I find my spirits and my health affect each other reciprocally—that is to say, every thing that discomposes my mind, produces a correspondent disorder in my body; and my bodily complaints are remarkably mitigated by those considerations that dissipate the clouds of mental chagrin." Cf. p. 393: "In less than a year, I make no doubt, but he will find himself perfectly at ease both in his mind and body, for the one had dangerously affected the other."

37   Bruce, *Radical Dr. Smollett*, p. 48.

38   Isaac Kramnick, *Republicanism and Bourgeois Radicalism: Political Ideology in Late Eighteenth-Century England and America* (Ithaca: Cornell University Press, 1990), p. 92.

39   Mary Wollstonecraft, *A Vindication of the Rights of Woman* (New York: Norton, 1988), p. 43.

40   Locke, *An Essay Concerning Human Understanding* (New York: Oxford University Press, 1979), p. 229.

41   H. W. Brands, *The First American: The Life and Times of Benjamin Franklin* (New York: Doubleday, 2000), pp. 631-32.

42   Benjamin Franklin, *Writings* (New York: Library of America, 1987), pp. 1188, 1225-26.

43   A version of this chapter appeared as "Imagination's Trickery: The Discovery of the Placebo Effect," *Journal of the Historical Society* 10 (2010): 57-73.

Chapter Seven

1   He is also a study in the tyranny of fantasy, somewhat like Burton's melancholics "molested by phantasy" and Dickensian grotesques ruled by fictions of their own invention.

2   According to Falconer, "it is but too usual with parents to foster the sensibility of their children, especially females, to an unusual degree, by officious attention to remove every thing that can give the least interruption to pleasure, or even awake the mind to its natural and necessary exertions." *A Dissertation on the Influence of the Passions Upon Disorders of the Body* (London, 1788), p. 75. Of hypochondria, which is roughly a male term for hysteria, Falconer writes, "The sufferers are mostly of gloomy disposition, and subject to a despondency of mind" (p. 59), a description that applies better to a melancholic than a man

like Mr. Woodhouse, said to be friendly and amiable. Falconer prescribes such social remedies for hypochondria as business, travel, diversion, and riding.

3   Falconer, *A Dissertation on the Influence of the Passions Upon Disorders of the Body*, pp. 72-73.

4   On the affectation of infirmity becoming real, see Montaigne's story of a certain man who hid from his Roman pursuers, assuming a disguise and pretending to be blind in one eye. "When he was able to recover a little liberty and wanted to rid himself of the plaster which he had worn so long over his eye, he found that he had actually lost the sight of that eye while under the mask. It is possible that his power of sight had been weakened by not having been exercised for such a long time. . ." "On Not Pretending to Be Ill" in Montaigne, *Complete Essays*, tr. M. A.Screech (London: Penguin, 2003), p. 781.

5   See Harold Merskey, *The Analysis of Hysteria, Second Edition: Understanding Conversion and Dissociation* (London: Gaskell, 1995), p. 194.

6   Cited in Jennifer Croswell, David Ransohoff, and Barnett Kramer, "Principles of Cancer Screening: Lessons from History and Study of Design Issues," *Seminars in Oncology*, June 2010. Doi:10.1053/j.seminoncol.2010.05.006: p. 10.

7   Robert Hahn, "The Nocebo Phenomenon: Scope and Foundations" in *The Placebo Effect: An Interdisciplinary Exploration*, ed. Anne Harrington (Cambridge, Mass.: Harvard University Press, 1997), p. 69.

8   Says Matthew Bramble, "I am persuaded that all valetudinarians are too sedentary, too regular, and too cautious—We should sometimes increase the motion of the machine, to *unclog the wheels of life*; and now and then take a plunge amidst the waves of excess, in order to case-harden the constitution. I have even found a change of company as necessary as a change of air, to promote a vigorous circulation of the spirits, which is the very essence and criterion of good health." *Humphry Clinker* (London: Penguin, 1985), pp. 381-82.

9   Falconer, *A Dissertation on the Influence of the Passions Upon Disorders of the Body*, pp. 71, 76.

10   Dr. Johnson's Rambler No. 4.

11   Peter Conrad, *The Medicalization of Society: On the Transformation of Human Conditions into Treatable Disorders* (Baltimore: Johns Hopkins University Press, 2007), p. 148.

12   On the lobbying campaign on behalf of multiple personality disorder, see Ian Hacking, *Mad Travelers: Reflections on the Reality of Transient Mental Illnesses* (Cambridge: Harvard University Press, 1998), p. 83. Conrad, *The Medicalization of Society*, discusses the cases of adult ADHD and chronic fatigue syndrome.

13   Elaine Showalter, *Hystories: Hysterical Epidemics and Modern Culture* (New York: Columbia University Press, 1997), p. 132.

Chapter Eight

1   *War and Peace*, tr. Louise and Aylmer Maude (New York: Oxford University Press, 1991), p. 617.

2   *War and Peace*, pp. 700-01. On Tolstoy and doctors see Harold Schefski, "Tolstoj's Case Against Doctors," *Slavic and East European Journal* 22 (1978): 569-73. On the unknowability of disease, cf. Montaigne, "On Experience" in *The Essays: A Selection*, tr. M. A. Screech (London: Penguin, 1993), p. 400: Nature "keeps her processes absolutely unknown. In her promises and threats there is great uncertainty, variability and obscurity." On the overweening "epistemic nihilism" implicit in Tolstoy's ridicule of the very possibility of medical knowledge, see Gary Saul Morson, *Hidden in Plain View: Narrative and Creative Potentials in 'War and Peace'* (Stanford: Stanford University Press, 1987), p. 172 and n.

3   If, as some believe, the reassuring sense of being in the hands of superior power can fuel the placebo effect, no wonder the doctors put on authoritarian airs in front of Ivan Ilych. Unfortunately, however, he is beyond the reach of the placebo effect.

4   Howard Spiro, "Clinical Reflections on the Placebo Phenomenon" in *The Placebo Effect*, ed. Anne Harrington (Cambridge, Mass.: Harvard University Press, 1997), p. 42.

5   All quotations are from the Maude translation in *The Portable Tolstoy*, ed. John Bayley (London: Penguin, 1978).

6   *Tolstoy's Short Fiction*, tr. Michael Katz (New York: Norton, 1991), p. 165.

7   Gary Saul Morson, *"Anna Karenina" in Our Time: Seeing More Wisely* (New Haven: Yale University Press, 2007), p. 10. In the case of Ivan Ilych, the sufferings of the body seem to give rise to even greater sufferings of the mind.

8   *War and Peace*, p. 702.

9   *The Brothers Karamazov*, tr. Constance Garnett (New York: Vintage, 1955), p. 281.

10   *What Is Art?*, tr. Richard Pevear and Larissa Volokhonsky (New York: Penguin, 1995), pp. 119-20.

Chapter Nine

1   The guilt that contributes to her grief might itself be reckoned a marker of depression.

2   Charlotte Blease, "Deception as Treatment: The Case of Depression," *Journal of Medical Ethics* online, Oct. 20, 2010. Doi: 10.1136/jme.2010.039313. Cognitive behavioral therapy is spoofed in David Lodge's *Therapy* (New York: Penguin, 1995).

3   Allan Horwitz and Jerome Wakefield, *The Loss of Sadness: How Psychiatry Transformed Normal Sorrow into Depressive Disorder* (New York: Oxford University Press, 2007), p. 20. Cf. Allan Horwitz, *Creating Mental Illness* (Chicago: University of Chicago Press, 2002), pp. 101-2: People in a state of depression "may realize, often from past experiences, that after these stressful experiences end, their distress will naturally dissipate over time. These lay views of distressing experiences may be more accurate than the illness models promoted by pharmaceutical companies and mental health professionals." For a fictional example of acute depression lasting but a few hours, see Chitra Divakaruni, "Mrs. Dutta Writes a Letter," reprinted and discussed in Bradley Lewis, *Narrative Psychiatry: How Stories Can Shape Clinical Practice* (Baltimore: Johns Hopkins University Press, 2011).

4   Shelley Taylor and Jonathon Brown, "Illusion and Well-Being: A Social Psychological Perspective on Mental Health," *Psychological Bulletin* 103 (1988): 193-210.

5   Taylor and Brown, "Illusion and Well-Being": 197.

6   Herbert Marcuse, *One-Dimensional Man: Studies in the Ideology of Advanced Industrial Society* (Boston: Beacon,1964), p. 84.

7   Jerome Frank and Julia Frank, *Persuasion and Healing: A Comparative Study of Psychotherapy* (Baltimore: Johns Hopkins University Press, 1991), e.g. p. 48: "To be effective, interpretations, the primary means of transmitting the therapist's conceptual framework, need not be correct, only plausible." Cf. p. 72: ". . . the criterion of the 'truth' of a psychotherapeutic interpretation would be one that is most satisfying or that makes the most sense to the particular person."

8   Frank and Frank, *Persuasion and Healing*, p. 151.

9   Sissela Bok, *Lying: Moral Choice in Public and Private Life* (New York: Vintage, 1978). For an example of explicitly paternalistic thinking about the use of placebos, see Franklin Miller and Luana Colloca, "The Placebo Phenomenon and Medical Ethics: Rethinking the Relationship Between Informed Consent and Risk-Benefit Analysis," *Theoretical Medicine and Bioethics* 32 (2011): 229-43. The authors espouse a "libertarian paternalist strategy of promoting

placebo and minimizing nocebo responses" (240)—an oxymoron.

10   Taylor and Brown, "Illusion and Well-Being": 204.

11   Taylor and Brown, "Illusion and Well-Being": 195.

12   When the storm sets in, Brekhunov declines the offer to stay the night in a neighboring village, as "four miles of good road, two of which lay through the forest, seemed easy to manage." Along with an inordinately high opinion of oneself, the cardinal delusions cited by Taylor and Brown are "exaggerated perception of control or mastery, and unrealistic optimism" (e.g., p. 193).

13   *The Praise of Folly and Other Writings*, ed. and tr. Robert M. Adams (New York: Norton, 1989), p. 39.

14   Francis Bacon, *The Major Works*, ed. Brian Vickers (New York: Oxford University Press, 2008), p. 341. In the "Digression Concerning the Original, the Use, and Improvement of Madness in a Commonwealth" (which sounds a good deal like a discourse on folly) in *A Tale of a Tub*, Swift's narrator lays it down that "if we take an examination of what is generally understood by *happiness*, as it has respect either to the understanding or the senses, we shall find all its properties and adjuncts will herd under this short definition: that *it is a perpetual possession of being well deceived*. And first, with relation to the mind or understanding, 'tis manifest what mighty advantages fiction has over truth; and the reason is just at our elbow, because imagination can build nobler scenes and produce more wonderful revolutions than fortune or nature will be at expense to furnish." Swift, *A Tale of a Tub and Other Works* (New York: Oxford University Press, 1986), p. 83.

15   Robert Burton *The Anatomy of Melancholy* (New York: New York Review Press, 2001), II.126-29 passim.

Chapter Ten

1   Allan Horwitz, *Creating Mental Illness* (Chicago: University of Chicago Press, 2002), p. 88. Among the categorical mental disorders is depression. Cf. p. 15: "A consequence of categorizing a broad scope of behavior as 'mental disorders' has been our considering much ordinary social behavior as pathological and overestimating the prevalence of mental disorder." Cf. Allan Horwitz and Jerome Wakefield, *The Loss of Sadness: How Psychiatry Transformed Normal Sorrow into Depressive Disorder* (New York: Oxford University Press, 2007). News stories playing into this trend are a genre. As I write (September 5, 2011), it is reported that the Centers for Disease Control and Prevention find "unac-

ceptably high levels of mental illness in the United States," with 25% of adults reporting such an illness in the preceding year and half of the population predicted to have one at some point.

2   Marcia Angell, "The Epidemic of Mental Illness: Why?", *New York Review of Books*, June 23, 2011.

3   Edward Shorter, *Doctors and Their Patients: A Social History* (New Brunswick: Transaction, 1991), p. 218.

4   Shorter, *Doctors and Their Patients*, p. 218.

5   B. Timothy Walsh et al., "Placebo Response in Studies of Major Depression," *JAMA* 287 (2002): 1840-47. On antidepressants' dependence on the placebo effect see Irving Kirsch, *The Emperor's New Drugs* (New York: Basic, 2010).

6   The drop-out rate for cognitive-behavioral therapy for depression is high, perhaps as high as 50%. David Mohr et al., "Effect of Telephone-Administered vs Face-to-face Cognitive Behavioral Therapy on Adherence to Therapy and Depression Outcomes Among Primary Care Patients," *JAMA* 307 (2012): 2278.

7   Andrew Lakoff, "The Right Patients for the Drug: Managing the Placebo Effect in Antidepressant Trials," *BioSocieties* 2 (2007): 64.

8   Bradley Lewis, *Narrative Psychiatry: How Stories Can Shape Clinical Practice* (Baltimore: Johns Hopkins University Press, 2011), p.1.

9   The same has been said of psychotherapy in general. "To a large extent, the demand for psychotherapy keeps pace with the supply, and one gets the uneasy feeling that the supply may even create the demand." Jerome Frank and Julia Frank, *Persuasion and Healing: A Comparative Study of Psychotherapy* (Baltimore: Johns Hopkins University Press, 1991), p. 9.

10   Frederick Crews, "Talking Back to Prozac," *New York Review of Books*, Dec. 6, 2007.

11   Horwitz and Wakefield, *Loss of Sadness*, p. 162.

12   A. Branthwaite and P. Cooper, "Analgesic Effects of Branding in Treatment of Headaches," *British Medical Journal* 282 (1981): 1576.

13   Walsh et al., "Placebo Response in Studies of Major Depression": 1844. My emphasis.

14   Some argue that chronic lymphocytic leukemia should not be called leukemia, "pointing out the harmful consequences of such a frightful label." Charles Bardes, *Pale Faces: The Masks of Anemia* (New York: Bellevue Literary Press, 2008), p. 102. On avoidance of the label "cancer," see Susan Sontag, *Illness as Metaphor and AIDS and Its Metaphors* (New York: Picador, 1989), pp. 6-7.

15   David Jopling, *Talking Cures and Placebo Effects* (New York: Oxford University Press, 2008), p. 111.

16   Irving Kirsch, "Antidepressants and the Placebo Response"; read in manuscript.

17   As Allan Horwitz makes clear, many mutually-reinforcing constituencies have a hand in promoting the notion that tens of millions of Americans suffer from mental disorders. The constituencies include pharmaceutical companies, clinicians, and "advocacy groups." See *Creating Mental Illness*, e.g., p. 106.

18   In the wonderful prologue to *The Arabian Nights* King Shahzaman is plunged into depression after suffering his wife's betrayal, only to recover when he discovers that his brother's wife is also dishonest. "He began to find consolation in his own affliction and forget his grief. When supper came, he ate and drank with relish and zest and, feeling better, kept eating and drinking, enjoying himself and feeling happy. He thought to himself, 'I am no longer alone in my misery; I am well.'" *Arabian Nights*, tr. Husain Haddawy (New York: Knopf, 1990), pp. 5-6.

19   Horwitz, *Creating Mental Illness*, p. 100.

20   Howard Spiro, *Doctors, Patients, and Placebos* (New Haven: Yale University Press, 1986), p. 3.

21   Horwitz and Wakefield, *Loss of Sadness*, e.g., pp. 26, 136, 138, 145.

22   Kirsch, *Emperor's New Drugs*, p. 104.

23   N. McKendrick, "Josiah Wedgwood: An Eighteenth-Century Entrepreneur in Salesmanship and Marketing Techniques," *Economic History Review*, New Series 12 (1960): 412.

24   Marcia Angell, "The Illusions of Psychiatry," *New York Review of Books*, July 14, 2011.

25   Benjamin Douglas Perkins, *The Influence of Metallic Tractors on the Human Body* (London: J. Johnson, 1798), p. 34.

26   Perkins, *The Influence of Metallic Tractors on the Human Body*, p. 86.

27   Spiro, *Doctors, Patients, and Placebos*, p. 1.

28   Kirsch, "Antidepressants and the Placebo Response," p. 15.

29   On Galvanism and Tractoration see John Greenway, "Galvanism as Therapeutic Agent: Perkins's 'Metallic Tractors' and the Placebo Effect," *ANQ* 14 (2001): 24-37.

30   Angell, "The Epidemic of Mental Illness: Why?"

31   Perkins, *The Influence of Metallic Tractors on the Human Body*, pp. 38, 40.

32   A feat also claimed by certain mesmerists. Robert Darnton, *Mesmerism and the End of the Enlightenment in France* (Cambridge, MA: Harvard University Press, 1968), p. 68.

33   Fabrizio Benedetti, *Placebo Effects: Understanding the Mechanisms in Health*

and Disease (New York: Oxford University Press, 2009), p. 125.

34   Irving Kirsch, "Specifying Nonspecifics: Psychological Mechanisms of Placebo Effects" in *The Placebo Effect*, ed. Anne Harrington (Cambridge, Mass.: Harvard University Press, 1997), p. 180.

35   John Haygarth, *Of the Imagination as a Cause and as a Cure of Disorders of the Body* (Bath: Cruttwell, 1800), p. 4.

36   Horwitz, *Creating Mental Illness*, p. 187.

37   Haygarth, *Of the Imagination as a Cause and as a Cure of Disorders of the Body*, p. 5.

38   Rich Mayes and Allan Horwitz, "DSM-III and the Revolution in the Classification of Mental Illness," *Journal of the History of the Behavioral Sciences* 4 (2005): 263.

39   Haygarth, *Of the Imagination as a Cause and as a Cure of Disorders of the Body*, p. 28.

40   On Prozac as a remedy for poor self-esteem, see, e.g., Samuel Barondes, "Thinking about Prozac," *Science*, Feb. 25, 1994: 1102.

41   Ulrich Tröhler, *"To Improve the Evidence of Medicine": The 18ᵗʰ Century British Origins of a Critical Approach* (Edinburgh: Royal College of Physicians, 2000), pp. 51-52, 93, 88.

42   Haygarth, *Of the Imagination as a Cause and as a Cure of Disorders of the Body,* p. 5.

43   Perkins, *The Influence of Metallic Tractors on the Human Body*, pp. 56-57, 69.

44   Arthur Shapiro and Elaine Shapiro, *The Powerful Placebo: From Ancient Priest to Modern Physician* (Baltimore: Johns Hopkins University Press, 1997), p. 95.

45   Cited in Gilbert Honigfeld, "Non-Specific Factors in Treatment," *Diseases of the Nervous System* 25 (1964): 146. The author is thought to be Armand Trousseau, who used placebos to test the efficacy of homeopathic treatments but believed in magnetic treatments (Shapiro and Shapiro, *Powerful Placebo*, p. 57).

46   See "The Depressing News About Antidepressants." See also "The Placebo Problem Big Pharma's Desperate to Solve," *Wired.co.uk*, 14 September 2009.

47   Barondes, "Thinking about Prozac": 1102.

48   David Wootton, *Bad Medicine: Doctors Doing Harm Since Hippocrates* (Oxford: Oxford University Press, 2006), p. 68.

Chapter Eleven

1   Francine Shapiro, *Getting Past Your Past: Taking Control of Your Life with Self-Help Techniques from EMDR Therapy* (New York: Rodale, 2012), p. 6.

2   Richard McNally, "EMDR and Mesmerism: A Comparative Historical Analysis," *Journal of Anxiety Disorders* 13 (1999): 225-36. See also James Herbert et al., "Science and Pseudoscience in the Development of Eye Movement Desensitization and Reprocessing: Implications for Clinical Psychology," *Clinical Psychology Review* 20 (2000): 945-71; and Gerald Rosen et al., "Power Therapies, Miraculous Claims, and Cures that Fail," *Behavioural and Cognitive Psychotherapy* 26 (1998): 99-101.

3   Francine Shapiro and Margot Silk Forrest, *EMDR: The Breakthrough Therapy for Overcoming Anxiety, Stress, and Trauma* (New York: Basic, 2004; orig. pub. 1997), p. 11. Subsequent page references are given in my text.

4   Robert Darnton, *Mesmerism and the End of the Enlightenment in France* (Cambridge: Harvard University Press, 1968), p. 4; Stephen Jay Gould, *Bully for Brontosaurus: Reflections on Natural History* (New York: Norton, 1991), p. 185; Henri F. Ellenberger, *The Discovery of the Unconscious: The History and Evolution of Dynamic Psychiatry* (New York: Basic, 1970), p. 62. On the later history of Mesmerism see Alison Winter, *Mesmerized: Powers of Mind in Victorian Britain* (Chicago: University of Chicago Press, 1998).

5   Darnton, *Mesmerism*, p. 59.

6   Jean Starobinski, *Jean-Jacques Rousseau: Transparency and Obstruction*, tr. Arthur Goldhammer (Chicago: University of Chicago Press, 1988).

7   Northrop Frye, *The Well-Tempered Critic* (Bloomington: Indiana University Press, 1963), pp. 42-43.

8   Roy Porter, "Mesmerism in England," *History Today* 35 (Sept. 1985): 23.

9   Darnton, *Mesmerism*, p. 50.

10   Jean-Jacques Rousseau, *The Confessions*, tr. J. M. Cohen (Harmondsworth, Middlesex: Penguin, 1953), p. 362.

11   Howard Spiro, "Clinical Reflections on the Placebo Phenomenon" in *The Placebo Effect: An Interdisciplinary Exploration*, ed. Anne Harrington (Cambridge: Harvard University Press, 1997), p. 42. Emphasis in the original.

12   See Ricky Greenwald, "The Power of Suggestion: Comment on *EMDR and Mesmerism: A Comparative Historical Analysis*," *Journal of Anxiety Disorders* 13 (1999): 611-15 and Richard McNally, "On Eye Movements and Animal Magnetism: A Reply to Greenwald's Defense of EMDR," *Journal of Anxiety Disorders* 13 (1999): 617-620.

13   Ellenberger, *Discovery of the Unconscious*, p. 63; Francine Shapiro, *Eye Movement Desensitization and Reprocessing: Basic Principles, Protocols, and Procedures* (New York: Guilford Press, 1995), p. 316.

14   On the utopian character of Mesmerism and the evolution of that movement into psychotherapy, see e.g. Fred Kaplan, *Dickens and Mesmerism: The Hidden Springs of Fiction* (Princeton: Princeton University Press, 1975).

15   Darnton, *Mesmerism*, p. 117.

16   McNally, "EMDR and Mesmerism": 230.

17   Shapiro, *Eye Movement Desensitization and Reprocessing: Basic Principles, Protocols, and Procedures*, p. 31.

18   Shapiro, *Eye Movement Desensitization and Reprocessing: Basic Principles, Protocols, and Procedures*, p. 40.

19   Louise Maxfield, "EMDR Milestones: The First 20 Years," *Journal of EMDR Practice and Research* 3 (2009): 211.

20   Darnton, *Mesmerism*, p. 8.

21   It is in keeping with the elementariness, or the literalism, of EMDR discourse that this "train of thought" is envisioned at one point as an actual train. "Ideally the person doing EMDR will feel as though she is on a train and the upsetting targeted events are merely the passing scenery." Shapiro and Forrest, *EMDR*, p. 52.

22   Shapiro, *Eye Movement Desensitization and Reprocessing: Basic Principles, Protocols, and Procedures*, p. 362.

23   Gary Saul Morson, *"Anna Karenina" in Our Time: Seeing More Wisely* (New Haven: Yale University Press, 2007), p. 120.

24   Lopez, "Franklin and Mesmer": 330.

25   Howard Brody and Daralyn Brody, *The Placebo Response: How You Can Release the Body's Inner Pharmacy for Better Health* (New York: HarperCollins, 2000). Offering ways "to let the inner pharmacy operate unimpeded" (p. 189), the authors premise their suggestions on the same metaphor of obstruction that powers EMDR.

26   Shapiro, *Eye Movement Desensitization and Reprocessing: Basic Principles, Protocols, and Procedures*, p. v.

27   See my *Fool's Paradise: The Unreal World of Pop Psychology* (Chicago: Ivan R. Dee, 2005).

28   Theodore Rubin, *Compassion and Self-Hate: An Alternative to Despair* (New York: Simon & Schuster, 1975), p. 170.

29   On the potential of placebo research to generate a paradigm-shift, see e.g. Karin Meissner, Niko Kohls, and Luana Colloca, "Introduction to Placebo

Effects in Medicine: Mechanisms and Clinical Implications," *Philosophical Transactions of the Royal Society: Biological Sciences* 366 (2011): 1787: "It is plausible to argue that research on placebo and nocebo effects may not only prompt a revolutionary shift in thinking of [sic] the physician-patient interaction, with the promise to guide strategies for optimizing clinical practice, but will also open promising avenues for improvement within most areas of modern medicine." Such rhetoric helps account for some of the excessive enthusiasm for placebos now in evidence.

30   See the discussion by Bruce Wampold and Joel Weinberger, "Critical Thinking and Psychotherapy Research" in *The Psychotherapy of Hope: The Legacy of Persuasion and Healing*, eds. Renato Alarcón and Julia Frank (Baltimore: Johns Hopkins University Press, 2012), pp. 12-13.

31   Richard Rechtman, "The Rebirth of PTSD: The Rise of a New Paradigm in Psychiatry," *Social Psychiatry and Psychiatric Epidemiology* 39 (2004): 913-14.

32   Porter, "Mesmerism in England": 22-30.

33   Jessica Riskin, *Science in the Age of Sensibility: The Sentimental Empiricists of the French Enlightenment* (Chicago: University of Chicago Press, 2002), pp. 192-93.

34   Richard McNally, "Progress and Controversy in the Study of Posttraumatic Stress Disorder," *Annual Review of Psychology* 54 (2003): 231-32.

35   McNally, "Progress and Controversy in the Study of Posttraumatic Stress Disorder": 230. Cf. Wilbur Scott, "PTSD in DSM-III: A Case in the Politics of Diagnosis and Disease," *Social Problems* 37 (1990): 294-310 and Allan Young, *The Harmony of Illusions: Inventing Post-Traumatic Stress Disorder* (Princeton: Princeton University Press, 1995), pp. 107-11. On the lobbying by members of the multiple personality movement to include the "multiple" diagnosis in *DSM-III*, see Ian Hacking, *Mad Travelers: Reflections on the Reality of Transient Mental Illness* (Cambridge, MA: Harvard University Press, 1998), p. 83. "Movement" is Hacking's term.

36   Rechtman, "The Rebirth of PTSD": 913-14.

37   Nancy Andreasen, "Acute and Delayed Posttraumatic Stress Disorders: A History and Some Issues," *American Journal of Psychiatry* 161 (2004): 1322. For an example of the take-off of PTSD, see I. Anastasiou et al., "Symptoms of Acute Posttraumatic Stress Disorder in Prostate Cancer Patients Following Radical Prostatectomy," *American Journal of Men's Health* 5 (2011): 84-89. Prostate cancer is recognized in the medical literature as an overdiagnosed disease. For the argument that "everyday life experiences" can produce PTSD, see Shapiro, *Getting Past Your Past*, p. 11.

38   A version of this chapter appeared in *Yale Journal of Biology and Medicine* 84
     (2011): 15-25.

Chapter Twelve

1   Arthur Shapiro and Elaine Shapiro, *The Powerful Placebo: From Ancient Priest
     to Modern Physician* (Baltimore: Johns Hopkins University Press, 1997), e. g.,
     p. 2.
2   Gilbert Honigfield, "Non-Specific Factors in Treatment," *Diseases of the Ner-
     vous System* 25 (1964): 145.
3   Anne Harrington, *The Placebo Effect: An Interdisciplinary Exploration* (Cam-
     bridge, MA: Harvard University Press, 1997), p. 1.
4   J. L. Mommaerts and Dirk Devroey, "The Placebo Effect: How the Subcon-
     scious Fits In," *Perspectives in Biology and Medicine* 55 (2012): 54. Cf. Howard
     Spiro, *The Power of Hope: A Doctor's Perspective* (New Haven: Yale University
     Press, 1998), p. 26: "Most physicians have grown defensive about [placebos],
     saying that their colleagues might use placebos but attributing their own thera-
     peutic success to less personal factors."
5   Damien Finniss et al., "Biological, Clinical, and Ethical Advances of Placebo
     Effects," *Lancet* 375 (2010): 692: "Recent data suggest that prescriptions of
     sugar pills and saline injections are rare, but that clinicians often prescribe vari-
     ous active treatments with the main intent of promoting a placebo response or
     complying with the wishes of patients."
6   Amir Raz et al., "Placebos in Clinical Practice: Comparing Attitudes, Beliefs,
     and Patterns of Use Between Academic Psychiatrists and Nonpsychiatrists,"
     Canadian Journal of Psychiatry 56 (2011): 198-208.
7   Veronica de Jong, "Active Expectations: Insights on the Prescription of Sub-
     Therapeutic Doses of Antidepressants for Depression," read in manuscript.
8   Irving Kirsch, *The Emperor's New Drugs: Exploding the Antidepressant Myth*
     (New York: Basic, 2010), pp. 157, 165.
9   Elaine Showalter, *Hysteries: Hysterical Epidemics and Modern Culture* (New
     York: Columbia University Press, 1997), p. 48. According to Allan Horwitz,
     Creating Mental Illness (Chicago: University of Chicago Press, 2002), p. 4, the
     number of mental-health professionals in the United States quadrupled form
     1970 to 1995.
10   Edward Shorter, *Doctors and Their Patients: A Social History* (New Bruns-
     wick: Transaction, 1991), p. 218.

11   Shapiro and Shapiro, *Powerful Placebo*, p. 231. One of the authors was himself a psychiatrist.

12   Fabrizio Benedetti, *Placebo Effects: Understanding the Mechanisms in Health and Disease* (New York: Oxford University Press, 2009), p. 142.

13   David Rosenthal and Jerome Frank, "Psychotherapy and the Placebo Effect," *Psychological Bulletin* 53 (1956): 294. Cf. Bruce Wampold and Joel Weinberger, "Critical Thinking and Psychotherapy Research" in *The Psychotherapy of Hope: The Legacy of Persuasion and Healing*, eds. Renato Alarcón and Julia Frank (Baltimore: Johns Hopkins University Press, 2012), pp. 4-5: "Frank and the Johns Hopkins group began investigating psychotherapy just as medicine was beginning to use placebo comparisons to control for various psychological factors, such as hope, expectancy, and the relationship with the physician."

14   Jerome Frank and Julia Frank, *Persuasion and Healing: A Comparative Study of Psychotherapy* (Baltimore: Johns Hopkins University Press, 1991), p. 48. Cf. David Jopling, *Talking Cures and Placebo Effects* (New York: Oxford University Press, 2008), pp. 44-48, 261-62.

15   Horwitz, *Creating Mental Illness*, p. 186.

16   E.g. Howard Spiro, *Doctors, Patients, and Placebos* (New Haven: Yale University Press, 1986), with Frank cited on p. 217. In this reflective book, critical thought ceases at the mention of psychotherapy. Psychotherapy is good. The possibility that the tenets and postulates of Freudian analysis might be elaborately fictitious, that the many modes of psychotherapy might offer openings for spurious insight, that therapeutic advice might not be harmless, that talk therapies might abuse the placebo effect, is simply not entertained.

17   Jerome Frank, "The Placebo Is Psychotherapy," *The Behavioral and Brain Sciences* 6 (1983): 292.

18   Bradley Lewis, *Narrative Psychiatry: How Stories Can Shape Clinical Practice* (Baltimore: Johns Hopkins University Press, 2011), p. 37; cf. pp. 148, 159.

19   Thomas Baskin et al., "Establishing Specificity in Psychotherapy: A Meta-Analysis of Structural Equivalence of Placebo Controls," *Journal of Consulting and Clinical Psychology* 71 (2003): 974.

20   Martin Seligman, "The Effectiveness of Psychotherapy: The Consumer Reports Study," *American Psychologist* 50 (1995): 965, 974. On the controversy surrounding Seligman's advocacy of Comprehensive Soldier Fitness—a psychological training program for the Army—without first testing it, see *The Chronicle of Higher Education*, Oct. 31, 2011.

21   Lloyd Wells and Julia Frank, "Psychodynamic Psychotherapy: From Psychoanalytic Arrogance to Evidence-Based Modesty" in Alarcón and Frank, *Psycho-*

therapy of Hope, p. 191.

22   Morris Parloff, "Placebo Controls in Psychotherapy Research: A Sine Qua Non or a Placebo for Research Problems?" *Journal of Consulting and Clinical Psychology* 54 (1986): 79-87.

23   Bruce Wampold et al., "The Story of Placebo Effects in Medicine: Evidence in Context," *Journal of Clinical Psychology* 63 (2007): 380-81.

24   Fabrizio Benedetti, *The Patient's Brain: The Neuroscience in the Doctor-Patient Relationship* (New York: Oxford University Press, 2011), p. 167. Cf. Horwitz, *Creating Mental Illness*, pp. 204-05.

25   Howard Brody, "The Placebo Response: Recent Research and Implications for Family Medicine," *Journal of Family Practice* 49 (2000): 649.

26   Marie Prévost, Anna Zuckerman, and Ian Gold, "Trust in Placebos," *Journal of Mind-Body Regulation* 1(2011): 139.

27   Shorter, *Doctors and Their Patients*, p. 160.

28   Shorter, *Doctors and Their Patients*, pp. 151, 210.

29   Cited in Honigfeld, "Non-Specific Factors in Treatment": 151.

30   Sissela Bok "The Ethics of Giving Placebos," *Scientific American*, November 1974: 17.

31   Barry Oken, "Placebo Effects: Clinical Aspects and Neurobiology," *Brain* 131 (2008): 2816.

32   Opher Caspi and Richard Bootzin, "Evaluating How Placebos Produce Change: Logical and Causal Traps and Understanding Cognitive Explanatory Mechanisms," *Evaluation & the Health Professions* 25 (2002): 452.

33   Franklin Miller and Ted Kaptchuk, "Deception of Subjects in Neuroscience: An Ethical Analysis," *Journal of Neuroscience* 28 (2008): 4841-43.

34   Herbert Benson and Richard Friedman. "Harnessing the Power of the Placebo Effect and Renaming It 'Remembered Wellness,'" *Annual Review of Medicine* 47 (1996): 193-99.

35   Klaus Linde, Margrit Fässler and Karin Meissner, "Placebo Interventions, Placebo Effects and Clinical Practice," *Philosophical Transactions of the Royal Society* B 366 (2011): 1906.

36   Sherwin Nuland, "Mind, Body, and the Doctor," *American Scholar* 70 (2001): 123-26.

37   Benedict Carey, "When Trust in Doctors Erodes, Other Treatments Fill the Void," *New York Times*, February 3, 2006. See also the report on alternative therapies, "The Believers," in *The Economist* of April 14, 2012.

38   Howard Spiro, "Clinical Reflections on the Placebo Phenomenon" in Harrington, *The Placebo Effect: An Interdisciplinary Exploration*, p. 39.

39   Shorter, *Doctors and Their Patients*, p. 195.

40   C. E. Kerr, I. Milner, and T. Kaptchuk, "William Cullen and a Missing Mind-Body Link in the Early History of Placebos," *Journal of the Royal Society of Medicine* 101 (2008): 89-92.

41   Shorter, *Doctors and Their Patients*, p. 157.

42   Seligman, "The Effectiveness of Psychotherapy": 976, 972.

43   Benedetti, *Placebo Effects*, p. 3.

44   Rachel Sherman and John Hickner, "Academic Physicians Use Placebos in Clinical Practice and Believe in the Mind-Body Connection," *Journal of General Internal Medicine* 23 (2008): 7.

45   Frank and Frank, *Persuasion and Healing*, p. 134.

46   Sherman and Hickner, "Academic Physicians Use Placebos": 8.

47   Jopling, *Talking Cures*, p. 47.

48   Jopling, *Talking Cures*, p. 31.

49   Gerald Koocher and Patricia Keith-Spiegel, *Ethics in Psychology and the Mental Health Professions* (New York: Oxford University Press, 2008), p. 102. Emphasis in the original.

50   Frank and Frank, *Persuasion and Healing*, p. 48.

51   Jopling, *Talking Cures*, p. 231.

52   Frank and Frank, *Persuasion and Healing*, pp. 146, 148.

53   Brody, "The Placebo Response: Recent Research and Implications for Family Medicine": 652.

54   Roger Rolls, *The Hospital of the Nation: The Story of Spa Medicine and the Mineral Water Hospital at Bath* (Bath: Avon, 1988), p. 166.

55   A kind of undeclared nostalgia for paternalism surfaces from time to time in placebo studies, as manifested in an emphasis on the physician who projects authority, makes confident predictions, and manipulates information in the patient's interest. On the last point see Franklin Miller and Luana Colloca, "The Placebo Phenomenon and Medical Ethics: Rethinking the Relationship between Informed Consent and Risk-Benefit Assessment," *Theoretical Medicine and Bioethics* 32 (2011): 229-43.

56   Austen, *Northanger Abbey*; Tolstoy, *Anna Karenina*. Shown reading a novel while under Vronsky's influence but before she closes with him, Anna Karenina identifies all too keenly with fictional others. "It was distasteful to her to read, that is, to follow the reflection of other people's lives. She had too great a desire to live herself. If she read that the heroine of the novel was nursing a sick man, she longed to move with noiseless steps about the room of the sick man; if she read of a member of Parliament making a speech, she longed to be delivering

the speech. . . . But there was no chance of doing anything." *Anna Karenina*, tr. Constance Garnett; rev. Leonard Kent and Nina Berberova (New York: Modern Library, 1993), p. 166.

57 Frank and Frank, *Persuasion and Healing*, p. 48. If talk therapies "provide the patient with both substitute relationships and a systematized Weltanschauung" as maintained by Herbert Adler and Van Buren Hammett, "The Doctor-Patient Relationship Revisited: An Analysis of the Placebo Effect," *Annals of Internal Medicine* 78 (1973): 597, a psychotherapeutic doctrine can easily appeal as plausible precisely because it affords a closed system of explanation.

58 In dismantling the now-fashionable cliché that we construct our life as a story, Galen Strawson observes that "It's well known that telling and retelling one's past leads to changes, smoothings, enhancements, shifts away from the facts. . . . The implication is plain: the more you recall, retell, narrate yourself, the further you are likely to move away from accurate self-understanding, from the truth of your being. Some are constantly telling their daily experiences to others in a storying way and with great gusto. They are drifting ever further from the truth." Strawson, "Against Narrativity" in *Real Materialism and Other Essays* (Oxford: Clarendon Press, 2008), p. 205. Cf. Gary Saul Morson, *"Anna Karenina"* in *Our Time: Seeing More Wisely* (New Haven: Yale University Press, 2007), p. 225: "The most meaningful moments of our lives do not fit life stories."

59 Spiro, *Power of Hope*, pp. 132-33.

60 Jopling, *Talking Cures*, p. 16. Cf. Wells and Frank, "Psychodynamic Psychotherapy," pp. 202-3: "Psychotherapeutic outcomes are not universally benign. Some people are clearly harmed by psychotherapy of many forms, including psychodynamic ones. Harm may reflect destructive views openly embraced by groups within the profession (e.g., the fairly recent epidemic of false memories of many kinds of abuse) . . ."

61 Frank and Frank, *Persuasion and Healing*, e.g., p. 72.

62 Frank and Frank, *Persuasion and Healing*, p. 72.

63 Showalter, *Hystories*, p. 59. Not necessarily Showalter's view.

64 Cf. Lewis, *Narrative Psychiatry*, p. 54: "For psychotherapy to be effective, the therapist and the client must have a sense of belief and confidence in their interpretive frames."

65 Jopling, *Talking Cures*, p. 257.

66 Jopling, *Talking Cures*, p. 258.

67 An earlier version of this chapter appeared as "From Medicine to Psychotherapy: The Placebo Effect," *History of the Human Sciences* 24 (2011): 95-107.

Chapter Thirteen

1   Communication from Jean-Luc Mommaerts. Cf. Jerome Frank and Julia Frank, *Persuasion and Healing: A Comparative Study of Psychotherapy* (Baltimore: Johns Hopkins University Press, 1993), p. 146.

2   Henry Beecher, "Ethics and Clinical Research," *New England Journal of Medicine* 274 (1966): 1354-60.

3   Franklin Miller and Ted Kaptchuk, "Deception of Subjects in Neuroscience: An Ethical Analysis," *Journal of Neuroscience* 28 (2008): 4841-43.

4   Cecilia Linde et al., "Placebo Effect of Pacemaker Implantation in Obstructive Hypertrophic Cardiomyopathy," *American Journal of Cardiology* 83 (1999): 903-07. Reportedly, study subjects gave informed consent (903).

5   For an opposing view see Irving Kirsch, *The Emperor's New Drugs: Exploding the Antidepressant Myth* (New York: Basic, 2010), p. 116. On "the experiences of friends and of other patients" feeding into our expectations concerning surgery, see Alan Johnson, "Surgery as a Placebo," *Lancet* 344 (1994): 1140.

6   Howard Brody and David Waters, "Diagnosis Is Treatment," *Journal of Family Practice* 10 (1980): 445-49.

7   Lee Park and Lino Covi, "Nonblind Placebo Trial: An Exploration of Neurotic Patients' Reponses to Placebo When Its Inert Nature Is Disclosed," *Archives of General Psychiatry* 12 (1965): 336-45. For a review and critique, see David Jopling, *Talking Cures and Placebo Effects* (Oxford: Oxford University Press, 2008), pp. 239-52.

8   Jopling, *Talking Cures and Placebo Effects*, p. 239.

9   Jopling, *Talking Cures and Placebo Effects*, p. 247.

10   Park and Covi, "Nonblind Placebo Trial": 337.

11   Fabrizio Benedetti et al., "Opioid-Mediated Placebo Responses Boost Pain Endurance and Physical Performance: Is It Doping in Sport Competitions?" *Journal of Neuroscience* 27 (2007): 11934-39.

12   Howard Spiro, *Doctors, Patients, and Placebos* (New Haven: Yale University Press, 1986), p. 3. Cf., however, Spiro, *The Power of Hope: A Doctor's Perspective* (New Haven: Yale University Press, 1998), p. 40: "Can you give yourself a placebo? . . . We can talk ourselves into hope, and do so every day."

13   Andrew Leuchter, "Scans Show How Placebo Aids Depression," CNN.com, Jan. 1, 2002. Confirmed to the author.

14   Cogent critics of deception in placebo research have argued that study subjects should be clearly informed that the trial they have entered will employ deception (and why); they do not argue that researchers should do away with decep-

tion by using open placebos. One can only assume that the reason they plead for "authorized deception" rather than the abolition of deception is that open placebos would defeat the aim of the study itself. Franklin Miller et al., "Deception in Research on the Placebo Effect," epub. Sept. 6, 2005, *PLoS Med*, e262. Cf. Miller and Kaptchuk, "Deception of Subjects in Neuroscience: An Ethical Analysis."

15   Amir Raz et al., "Placebos in Clinical Practice: Comparing Attitudes, Beliefs, and Patterns of Use Between Academic Psychiatrists and Nonpsychiatrists," *Canadian Journal of Psychiatry* 56 (2011): 199.

16   "The dominant view among medical researchers and clinicians deems placebo administration ethically problematic." Raz et al., "Placebos in Clinical Practice": 204.

17   A. Sandler and J. Bodfish, "Open-label Use of Placebos in the Treatment of ADHD: a Pilot Study," *Child: Care, Health and Development* 34 (2008): 104-10.

18   Adrian Sandler, Corinne Glesne, and James Bodfish, "Conditioned Placebo Dose Reduction: A New Treatment in ADHD?", *Journal of Developmental and Behavioral Pediatrics* 31 (2010): 369-75.

19   A. Sandler, C. Glesne, and G. Geller, "Children's and Parents' Perspectives on Open-Label Use of Placebos in the Treatment of ADHD," *Child: Care, Health and Development* 34 (2008): 120.

20   S. Karen Chung et al., "Revelation of a Personal Placebo Response: its Effect on Mood, Attitudes and Future Placebo Responding." *Pain* 132 (2007): 281-88.

21   For a defense of the use of placebos in clinical practice that gets around informed consent by playing on the equivocation of "implicit consent," see Bennett Foddy, "The Ethical Placebo," *Journal of Mind-Body Regulation* 1 (2011): 53-62.

22   Lene Vase et al., "The Contributions of Suggestion, Desire, and Expectation to Placebo Effects in Irritable Bowel Syndrome Patients: an Empirical Investigation," *Pain* 105 (2003): 17-25. Reportedly, subjects in a clinical trial "sometimes feel fortunate to be randomized to the placebo group because they believe this 'placebo' is a therapy they would not have received if they had not participated in the clinical trial." Kristin Mattocks and Ralph Horwitz, "Placebos, Active Control Groups, and the Unpredictability Paradox," *Biological Psychiatry* 47 (2000): 693.

23   Foddy, "The Ethical Placebo": 60.

24   Mary Crenshaw Rawlinson, "Truth-Telling and Paternalism in the Clinic: Philosophical Reflections on the Use of Placebos in Medical Practice" in *Pla-*

cebo: Theory, Research, and Mechanisms, ed. Leonard White, Bernard Tursky, and Gary Schwartz (New York: Guilford Press, 1985), pp. 410-11. Emphasis in the original.

25   T. Kaptchuk et al., "Placebos without Deception: A Randomized Controlled Trial in Irritable Bowel Syndrome," *PLoS ONE* 5 (2010): e15591. The authors also note that "it is likely our study . . . benefited from ongoing media attention giving credence to powerful placebo effects."

26   More, *Utopia*, tr. George Logan and Robert M. Adams (Cambridge: Cambridge University Press, 1989), p. 60.

27   Brody and Waters, "Diagnosis Is Treatment": 448-49. The risks of conferring a mantle of distress that society will accept are illustrated in Mr. Woodhouse.

28   Ted Kaptchuk et al., "Components of Placebo Effect: Randomised Controlled Trial in Patients with Irritable Bowel Syndrome," *BMJ*, doi:10.1136/bmj.39524.439618.25.

29   Herbert Adler and Van Buren Hammett, "The Doctor-Patient Relationship Revisited: An Analysis of the Placebo Effect," *Annals of Internal Medicine* 78 (1973): 598.

30   Sandler, Glesne, and Geller, "Children's and Parents' Perspectives on Open-Label Use of Placebos in the Treatment of ADHD": 118.

31   James House et al., "Social Relationships and Health," *Science* 241 (1988): 543.

32   Barron Lerner, *The Breast Cancer Wars: Hope, Fear, and the Pursuit of a Cure in Twentieth-Century America* (Oxford: Oxford University Press, 2001), p. 272.

33   David Spiegel et al., "Effects of Supportive-Expressive Group Therapy on Survival of Patients with Metastatic Breast Cancer: A Randomized Prospective Trial," *Cancer* 110 (2007): 1130.

34   Spiegel et al., "Effects of Supportive-Expressive Group Therapy on Survival of Patients with Metastatic Breast Cancer." This ironic coda is missing from the account of the Spiegel experiment given in Howard Brody and Daralyn Brody, *The Placebo Response: How You Can Release the Body's Inner Pharmacy for Better Health* (New York: HarperCollins, 2000), pp. 203-05. Similarly, while the Prostate Cancer Lifestyle Trial produced little of medical significance, some saw it as a vindication of the uplifting effects of groups all the same. "A sense of belonging to something greater, along with all the lifestyle program components, creates a synergy that helps men [in group sessions] feel positive and optimistic. . . . There is less conflict in their lives and they value a sense of community." C. Kronenwetter et al., "A Qualitative Analysis of Interviews of Men With Early Stage Prostate Cancer," *Cancer Nursing* 28 (2005): 106.

35   An earlier version of this chapter appeared in *Skeptic* 16: 2 (2011): 41-44.

Chapter Fourteen

1   Elaine Showalter, *Hystories: Hysterical Epidemics and Modern Culture* (New York: Columbia University Press, 1997), p. 124. CFS is known in Britain as ME—myalgic encephalomyelitis.

2   Howard Spiro, *The Power of Hope: A Doctor's Perspective* (New Haven: Yale University Press, 1998), p. 251.

3   Showalter, *Hystories.*

4   "Man has no Body distinct from his Soul," declares Blake in *The Marriage of Heaven and Hell.* On Ginsberg and Blake, see e.g. Frank Kermode, *The Sense of an Ending* (Oxford: Oxford University Press, 1966), p. 119.

5   Stanley Jackson, *Melancholia and Depression: From Hippocratic Times to Modern Times* (New Haven: Yale University Press, 1986), pp. 101-03. The "great suicidal dramas on the apartment cliff-banks of the Hudson" vaguely evoke *Hamlet,* set in a castle perched on a cliff that induces thoughts of suicide (1.4).

6   Ginsberg used to say that only on its surface was "Howl" a litany of suicides, by which he meant that it is really a poem of affirmation.

7   One of Ginsberg's heroes, Blake, famously said that Milton was of the devil's party without knowing it.

8   A. Alvarez, *The Savage God: A Study of Suicide* (New York: Norton, 1990), p. 231.

9   Lydia Ginzburg, *On Psychological Prose,* tr. Judson Rosengrant (Princeton: Princeton University Press, 1991), p. 38.

10   Arnold Stein, *Answerable Style: Essays on Paradise Lost* (Minneapolis: University of Minnesota Press, 1953), p. 75.

11   Timothy Jones et al., "Mass Psychogenic Illness Attributed to Toxic Exposure at a High School," *New England Journal of Medicine* 342 (2000): 96-100.

12   See Kermode, *Sense of an Ending,* p. 119.

13   Jonah Raskin, *American Scream: Allen Ginsberg's* Howl *and the Making of the Beat Generation* (Berkeley: University of California Press, 2004), pp. 144-45.

14   The society around us causes us to be sick. Thus the countercultural doctrine of the 1960s: "You are what you think: so you had better select your ideas with the utmost care. Ideas have medical consequences. Sick thoughts sicken." Geoffrey O'Brien, *Dream Time: Chapters from the Sixties* (Washington, D.C.: Counterpoint, 2002), p. 101.

15   Raskin, *American Scream,* p. 135.

16   Showalter, *Hystories,* p. 3.

Chapter Fifteen

1   Allan Horwitz and Jerome Wakefield, *The Loss of Sadness: How Psychiatry Transformed Normal Sorrow into Depressive Disorder* (New York: Oxford University Press, 2007), p. 5.

2   Mass screening for prostate cancer was introduced in the late 1980s, Prozac in 1987.

3   H. Gilbert Welch, Lisa Schwartz and Steven Woloshin, "What's Making Us Sick Is an Epidemic of Diagnoses," *New York Times*, Jan. 2, 2007.

4   Daniel Moerman, *Meaning, Medicine, and the "Placebo Effect"* (Cambridge: Cambridge University Press, 2002), p. 20. Would we allow the manufacturer to make such a claim?

5   Jennifer Croswell, David Ransohoff, and Barnett Kramer, "Principles of Cancer Screening: Lessons from History and Study of Design Issues," *Seminars in Oncology*, June 2010. Doi:10.1053/j.seminoncol.2010.05.006: p. 6. The case has also been made that the experience of seeing men through the terminal stages of prostate cancer can lead urologists to overestimate the value, and write off the harms, of screening the general population for the disease. See Barnett Kramer, "The Science of Early Detection," *Urological Oncology* 22 (2004): 344-47. Cf. Barnett Kramer and J. Miller Croswell, "Cancer Screening: The Clash of Science and Intuition," *Annual Reviews of Medicine* 60 (2009): 135: "Powerful, pervasive biases make reliance on experience alone a dangerous strategy." According to an article in the New York Times of Oct. 5, 2011 ("Can Cancer Ever Be Ignored?"), "The popularity of the P.S.A. test as the main weapon against prostate cancer is due in large measure to the earnest and passionate advocacy of William Catalona, a urologist from Northwestern University Feinberg School of Medicine," who became an early champion of the test as a direct result of seeing patients die.

6   Cf. H. Gilbert Welch, Lisa Schwartz, and Steven Woloshin, *Overdiagnosed: Making People Sick in the Pursuit of Health* (Boston: Beacon, 2011), pp. 187-88.

7   If low-grade prostate cancer were called something other than cancer as some urologists propose, it might not excite such alarm. Just as the wording of informed-consent documents can influence the occurrence of side effects, just as changing the name of a placebo makes a patient's response to it less predictable—two findings presented at a conference on Placebos in the Clinic (Montreal, 2012)—in the case of prostate cancer, too, language matters.

8   See e.g. Steven Woolf, "Screening for Prostate Cancer with Prostate-Specific

Antigen: An Examination of the Evidence," *New England Journal of Medicine* 333 (1995): 1401-05.

9   Michael Brawer and Paul Lange, "Prostate-Specific Antigen and Premalignant Change: Implications for Early Detection," *CA Cancer J Clin* 39 (1989): 373.

10   Nina Sharifi, and Barnett Kramer, "Screening for Prostate Cancer: Current Status and Future Prospects," *American Journal of Medicine* 120 (2007): 743. In 1995 it was estimated that 9 million American men harbored latent prostate cancer: Woolf , "Screening for Prostate Cancer": 1401.

11   Ian Thompson et al, "The Influence of Finasteride on the Development of Prostate Cancer," *New England Journal of Medicine* 349 (2003): 215-24.

12   "The likelihood of being diagnosed with [prostate cancer] is directly related to the rigor with which one looks for it." Howard Parnes, Ian Thompson, and Leslie Ford, "Prevention of Hormone-Related Cancers: Prostate Cancer." *Journal of Clinical Oncology* 23 (2005): 374.

13   See the Briefing Document prepared for the meeting of the FDA's Oncologic Drugs Advisory Committee, Dec.1, 2010: http://www.fda.gov/downloads/AdvisoryCommittees/CommitteesMeetingMaterials/Drugs/OncologicDrugsAdvisoryCommittee/UCM234934.pdf

14   Chris Magee and Ian Thompson, "Evidence of Effectiveness of Prostate Cancer Screening," in *Prostate Cancer Screening*, eds. Ian Thompson, Martin Resnick, and Eric Klein (Totowa, NJ: Humana Press, 2001), pp. 157-74. Cf. Laura Esserman, Yiwey Shieh, and Ian Thompson, "Rethinking Screening for Breast Cancer and Prostate Cancer. *JAMA* 302 (2009: 1685-92.

15   Evidence for the uncertain mortality benefits of PSA testing can be found in Gerald Andriole et al., "Mortality Results from a Prostate-Cancer Screening Trial, *New England Journal of Medicine* 360 (2009):1310-19; and Fritz Schröder et al., "Screening and Prostate-Cancer Mortality in a Randomized European Study," *New England Journal of Medicine* 360 (2009): 1320-28. Cf. the ERSPC update of March 2012. Moreover, disease mortality is to be distinguished from overall mortality, which has shown no decrease in trials of either breast- or prostate-cancer screening. "Inadequate statistical power to detect all-cause mortality reductions in nearly half a million women for mammography and nearly a quarter million men for PSA indicates that if there are all-cause mortality benefits from these modalities, they are extremely small, which belies widespread perceptions of breast and prostate cancer screening," David Newman, "Screening for Breast and Prostate Cancers: Moving Toward Transparency," *JNCI* 102 (2010): 1009.

16   On breast cancer medicine and the rhetoric of early detection, see Barron

Lerner, *The Breast Cancer Wars: Hope, Fear, and the Pursuit of a Cure in Twen-tieth-Century America* (Oxford: Oxford University Press, 2001). However, the intuitive appeal of early detection precedes the twentieth century. Five hundred years ago Machiavelli wrote, "What doctors say about consumption applies [in politics]: in the early stage it is hard to recognize and easy to cure, but in the later stages, if you have done nothing about it, it becomes easy to recognize and hard to cure." *The Prince*, tr. Robert M. Adams (New York: Norton, 1977), p. 9.

17   David Ransohoff et al., "Why Is Prostate Cancer Screening So Common When the Evidence Is So Uncertain? A System Without Negative Feedback," *American Journal of Medicine* 113 (2002): 663-67.

18   A PSA test yields either an unsuspicious or a suspicious result. If the for-mer, the tested man is relieved; if the latter, he is biopsied and cancer is either confirmed, in which case he is dismayed but presumably relieved that it was dis-covered early, or not confirmed, in which case he is again relieved. PSA testing seems to generate no outcome that does not reinforce the testing regime itself.

19   Welch, Schwartz, and Woloshin, *Overdiagnosed,* p. 175.

20   Thomas More, *Utopia*, tr. George Logan and Robert M. Adams (Cambridge: Cambridge University Press, 1989), p. 75.

21   Diane Fink, "Community Programs: Breast Cancer Awareness," *Cancer* 64 (1989): 2674-81.

22   Fink, "Community Programs: Breast Cancer Awareness."

23   E. David Crawford, "Prostate Cancer Awareness Week: September 22 to 28, 1997," *CA Cancer J Clin* 47 (1997): 288-96.

24   Zinelabidine Abouelfadel and E. David Crawford, "Experience of Prostate Cancer Awareness Week." In Ian Thompson, Martin Resnick, and Eric Klein, *Prostate Cancer Screening* (Totowa, NJ: Humana Press, 2001), p. 241.

25   Barbara Rimer, "Putting the 'Informed' in Informed Consent about Mammog-raphy," *JNCI* 87 (1995): 703-4.

26   Gerd Gigerenzer, Jutta Mata, and Ronald Frank "Public Knowledge of Benefits of Breast and Prostate Cancer Screening in Europe," *JNCI* 101 (2009): 1216-20.

27   Richard Hoffman et al., "Prostate Cancer Screening Decisions: Results from the National Survey of Medical Decisions (DECISIONS Study)," *Archives of Internal Medicine* 169 (2009): 1611-18. Cf. Alexandra Barratt et al., "Use of Decision Aids to Support Informed Choices about Screening," *BMJ* 329 (2004): 507-10. It is now generally recognized that the PSA revolution breached in-formed consent.

28   It was announced by the Radiotherapy Clinics of Georgia on Feb. 14, 2012

that the state of Georgia was issuing Prostate Cancer Awareness license plates.

29   Elaine Showalter, *Hystories: Hysterical Epidemics and Modern Media* (New York: Columbia University Press, 1997), pp. 4-5. Among the hysterias discussed are chronic fatigue syndrome and the recovered memory movement.

30   For an earlier version of this chapter see my "How Did the PSA System Arise?", *Journal of the Royal Society of Medicine* 103 (2010): 309-12.

Epilogue

1   Michel de Montaigne, *Apology for Raymond Sebond* (Indianapolis: Hackett, 2003), p. 150.

2   Luana Colloca and Fabrizio Benedetti, "Placebo Analgesia Induced by Social Observational Learning," *Pain* 144 (2009): 28-34.

3   Franklin Miller and Luana Colloca, "The Placebo Phenomenon and Medical Ethics: Rethinking the Relationship between Informed Consent and Risk-Benefit Assessment," *Theoretical Medicine and Bioethics* 32 (2011): 234.

4   Howard Spiro, *The Power of Hope: A Doctor's Perspective* (New Haven: Yale University Press, 1998), p. 241.

5   C. Kadoch et al., "When Brachytherapy Fails: A Case Report and Discussion," *The Oncologist* 10 (2005): 799-805.

6   Stewart Justman, "What's Wrong with Chemoprevention of Prostate Cancer?", *American Journal of Bioethics*, December 2011.

7   Howard Spiro, *Doctors, Patients, and Placebos* (New Haven: Yale University Press, 1986), p. 3.

Index

www.ingramcontent.com/pod-product-compliance
Lightning Source LLC
Chambersburg PA
CBHW031252090426
42742CB00007B/418